THE ANCIENT BRIDGES OF MID AND EASTERN ENGLAND

By

E. JERVOISE, A.M.Inst.C.E.

*Author of " The Ancient Bridges of the South of England ";
" The Ancient Bridges of the North of England "*

Written on behalf of

THE SOCIETY FOR THE PROTECTION
OF ANCIENT BUILDINGS

Architecture

Architecture (from the Latin *architectura*, after the Greek *arkhitekton*, meaning chief builder) is both the process and the product of planning, designing, and constructing buildings and other physical structures. It is an incredibly important part of human existence – starting from the simplest aspects of survival, yet also functioning as a cultural symbol, a works of art, and as a means of identification of past civilisations.

Building first evolved out of the dynamics between needs (shelter, security, worship, etc.) and means (available building materials and attendant skills). As human cultures developed and knowledge began to be formalized through oral traditions and practices, building became a craft, and 'architecture' was the formalised version of this craft. In many ancient civilizations, such as those of Egypt and Mesopotamia, architecture and urbanism reflected the constant engagement with the divine and the supernatural. Many ancient cultures resorted to monumentality in architecture (think of the Pyramids at Giza, or the Parthenon at Athens) to represent symbolically the political power of the ruler, the ruling elite, or the state itself.

The architecture and urbanism of the Classical civilizations such as the Greeks and the Romans generally evolved from civic ideals rather than religious or empirical ones – and new building types emerged. Architectural 'style' developed in the form of the Classical orders. The earliest surviving written work on the subject of architecture is *De Architectura*, by the Roman architect Vitruvius in the early first-century CE. According to Vitruvius, a good building should satisfy the three principles of *firmitas, utilitas,* and *venustas*, translating as 'durability', 'utility' and 'beauty'.

Early Asian writings on architecture include the *Kao Gong Ji* of China from the seventh century BCE; the *Shilpa Shastras* of ancient India and the *Manjusri Vasthu Vidya Sastra* of Sri Lanka. The architecture of different parts of Asia developed along different lines from that of Europe; Buddhist, Hindu and Sikh architecture each having different characteristics. Islamic architecture began in the seventh century CE, incorporating architectural forms from the ancient Middle East and Byzantium, but also developing features to suit the religious and social needs of the society. In Europe during the Medieval period, guilds were formed by craftsmen to organize their trades and written contracts have survived, particularly in relation to ecclesiastical buildings. From about 900 CE onwards, the movements of both clerics and tradesmen carried architectural knowledge across Europe, resulting in the pan-European styles Romanesque and Gothic.

In Renaissance Europe, from about 1400 onwards, there was a revival of Classical learning accompanied by the development of Renaissance Humanism, which placed greater emphasis on the role of the individual in society. Buildings were ascribed to specific architects – Brunelleschi, Alberti, Michelangelo, Palladio – and the cult of the individual had begun. Leone Battista Alberti, who elaborates on the ideas of Vitruvius in his treatise, *De Re Aedificatoria*, saw beauty primarily as a matter of proportion, although ornament also played a part. For Alberti, the rules of proportion were those that governed the idealised human figure; 'the Golden mean'.

The notion of 'style' in the arts was not developed until the sixteenth century, with the writing of Vasari. By the eighteenth century, his *Lives of the Most Excellent Painters, Sculptors, and Architects* had been translated into Italian, French, Spanish and English. With the emerging knowledge in scientific fields and the rise of new materials and technology, architecture and engineering further began to separate, and the architect began to concentrate on aesthetics and the humanist aspects, often at the expense of technical aspects of building design. Around this time, there was also the rise of the 'gentleman architect' who usually concentrated on visual qualities derived from historical prototypes, typified by the many country houses of Great Britain that were created in the Neo Gothic or Scottish Baronial styles.

The nineteenth-century English art critic, John Ruskin, in his *Seven Lamps of Architecture* (published 1849), had a representative view of what constituted architecture. Architecture was the 'art which so disposes and adorns the edifices raised by men ... that the sight of them contributes to his mental health, power, and pleasure.' For Ruskin, the aesthetic was of overriding significance. His work goes on to state that a building is not truly a work of architecture unless it is in some way 'adorned'. Around the beginning of the twentieth century, a general dissatisfaction with the emphasis on revivalist architecture and elaborate decoration gave rise to many new lines of thought that served as precursors to Modern Architecture.

Notable among these schools is the *Deutscher Werkbund*, formed in 1907 to produce better quality machine made objects. Following this lead, the *Bauhaus school*, founded in Weimar in 1919, redefined the architectural bounds; viewing the creation of a building as the ultimate synthesis – the apex of art, craft, and technology. When Modern architecture was first practiced, it was an avant-garde movement with moral, philosophical, and aesthetic underpinnings. Immediately after World War I, pioneering modernist architects sought to develop a completely new style appropriate for a new post-war social and economic order, focused on meeting the needs of the middle and working classes.

On the difference between the ideals of architecture and mere construction, the renowned twentieth-century architect Le Corbusier wrote:

> You employ stone, wood, and concrete, and with these materials you build houses and palaces: that is construction. Ingenuity is at work. But suddenly you touch my heart, you do me good. I am happy and I say: This is beautiful. That is Architecture.

Architecture itself has an incredibly long and fascinating history. As long as humans have been around, we have needed places to live, and have sought ways to make these spaces beautiful and functional. As our societies continue to change, so does the architecture which reflects them. It is hoped that the current reader enjoys this book on the subject.

FOREWORD

THIS volume, the third of the series, deals with the rivers and bridges in the eastern half of the district which is bounded by the counties of Lancaster and York on the North and by the river Thames on the South. It embraces the counties of Bedford, Buckingham, Cambridge, Derby, Essex, Hertford, Huntingdon, Leicester, Lincoln, Middlesex, Norfolk, Northampton, Nottingham, Oxford, Rutland, Stafford, and Suffolk, also the part of Warwickshire which is drained by tributaries of the river Trent.

The remaining rivers of the county of Warwick flow westwards into the river Severn and will be included in the fourth volume, dealing with Wales and Western England, which it is hoped may be published at a later date when the Survey of Ancient Bridges has been completed.

As explained in the Foreword of an earlier volume, endeavours have been made to visit as far as possible all crossings shown on the large-scale maps which were published towards the end of the eighteenth century. In addition, pack-horse bridges were examined wherever possible, but this book does not pretend to be a complete record of these, as many could only have been reached on a walking tour, and this Survey is concerned primarily with bridges which might be threatened by road-widening schemes. The Survey of the district covered by the present volume took several years to complete, and it is possible that some of the bridges described in this volume may since have been destroyed, but it is sincerely hoped that this is not the case.

In compiling the historical notes reference was made to the Publications of the Archæological and Record Societies of the Counties mentioned above and to many County histories. The Patent Rolls gave much useful information, and Leland's *Itinerary*, made in the reign of Henry VIII, and Ogilby's *Road Book*, compiled towards the end of the seventeenth century, helped to determine the dates of many bridges.

Much valuable history was found in the Records of the Quarter Sessions, which have fortunately been published for several counties, and the notes used in the case of Staffordshire were kindly given me by Mrs. A. L. Thomas, of Stoke, who had permission from Mr. Eustace Joy, Clerk of the Peace for Stafford, to make researches into the county Records.

Reviewers of the earlier volumes of this series have expressed the wish that maps might be included to show the position of the bridges described, but unfortunately this is quite impossible in a book published at the present price. It was suggested in the case of the second volume that a map should be included to show the courses of the various rivers, together with the chief towns, but it was finally decided that this would be of comparatively little value to the reader, as it would not have been possible to show the position of more than a few of the 440 bridges mentioned in that volume. The present book covers an even larger district.

E. J.

September, 1932.

CONTENTS

CHAPTER		PAGES
	FOREWORD	v
	LIST OF ILLUSTRATIONS	ix
I.	THE RIVER TRENT	1
II.	THE NORTHERN TRIBUTARIES OF THE TRENT	16
III.	THE SOUTHERN TRIBUTARIES OF THE TRENT	37
IV.	THE RIVERS AND BRIDGES OF LINCOLNSHIRE AND RUTLANDSHIRE	55
V.	THE RIVER NENE	71
VI.	THE RIVER OUSE	82
VII.	THE RIVERS AND BRIDGES OF EAST ANGLIA AND ESSEX	109
VIII.	THE NORTHERN TRIBUTARIES OF THE THAMES	136
	INDEX OF BRIDGES	161

LIST OF ILLUSTRATIONS

Huntingdon Bridge — *Frontispiece*

FIG.		FACING PAGE
1.	Sandon Bridge	2
2.	Essex Bridge, Great Haywood	3
3.	Essex Bridge, Great Haywood	3
4.	Swarkeston Bridge	14
5.	Cavendish Bridge	14
6.	Okeover Bridge	15
7.	Sandy-brook Bridge	15
8.	Hanging Bridge, Mayfield	20
9.	Ellaston Bridge	20
10.	Dove Bridge, Doveridge	21
11.	Monk's Bridge, Eggington	22
12.	Yorkshire Bridge	23
13.	Froggatt Bridge	23
14.	Baslow Bridge	26
15.	One Arch Bridge, Beeley	26
16.	Sheepwash Bridge, Ashford	27
17.	Bakewell Bridge	27
18.	Darley Bridge	28
19.	Cromford Bridge	28
20.	Merriel Bridge	29
21.	New Bridge, Blyth	29
22.	Water Orton Bridge	38
23.	Coleshill Bridge	38
24.	Hemlingford Bridge	39

LIST OF ILLUSTRATIONS

FIG.   FACING PAGE

25. Elford Bridge - - - - - 39
26. Enderby Mill Bridge - - - - 42
27. Old Bridge, Belgrave - - - - 43
28. Rearsby Bridge - - - - - 43
29. Anstey Bridge - - - - - 50
30. Fleming's Bridge, Bottesford - - - 50
31. Wakerley Bridge - - - - - 51
32. Duddington Bridge - - - - 51
33. Ketton Bridge - - - - - 62
34. Church Bridge, Empingham - - - 62
35. Deeping Gate Bridge - - - - 63
36. Trinity Bridge, Crowland - - - 63
37. Everdon Bridge - - - - - 70
38. Desborough Bridge - - - - 70
39. Geddington Bridge - - - - 71
40. Ditchford Bridge - - - - - 71
41. Irthlingborough Bridge - - - - 74
42. Brigstock Bridge - - - - - 74
43. Fotheringhay Bridge - - - - 75
44. Wansford Bridge - - - - - 75
45. Thornborough Bridge - - - - 84
46. Harrold Bridge - - - - - 84
47. Stafford Bridge - - - - - 85
48. Barford Bridge - - - - - 85
49. St. Neots Bridge - - - - - 92
50. Nuns' Bridge - - - - - 92
51. Alconbury Bridge - - - - 93
52. Spaldwick Bridge - - - - 93

LIST OF ILLUSTRATIONS

FIG.		FACING PAGE
53.	St. Ives Bridge	102
54.	Little Chesterford Bridge	102
55.	Abbot's Bridge, Bury St. Edmunds	103
56.	Moulton Bridge	103
57.	Brandon Bridge	106
58.	Houghton St. Giles (Foot) Bridge	106
59.	Wiveton Bridge	107
60.	Mayton Bridge	116
61.	Wey Bridge	117
62.	Heigham Bridge	117
63.	Cringleford Bridge	124
64.	Newton Flotman Bridge	125
65.	Bishop Bridge, Norwich	125
66.	Blyford Bridge	128
67.	Ash Street Bridge	128
68.	Toppesfield Bridge, Hadleigh	129
69.	Cattawade Bridge	144
70.	Long Bridge, Coggeshall	144
71.	Moulsham Bridge, Chelmesford	145
72.	Latchleys Manor House Bridge	145
73.	Woodford Bridge	152
74.	Watton at Stone Bridge	152
75.	Wadesmill Bridge	153
76.	Waterend Bridge	153
77.	Bourton Bridge	156
78.	Burford Bridge	156
79.	Eastleach Martin (Clapper) Bridge	157
80.	Bibury (Foot) Bridge	157

THE ANCIENT BRIDGES OF MID AND EASTERN ENGLAND

CHAPTER I

THE RIVER TRENT

NUMEROUS small streams, rising in the extreme north-west of Staffordshire, form the source of this important river, but no ancient bridges now remain on its upper reaches. It is interesting, however, to find the record of a bridge at Newcastle-under-Lyme as early as the reign of Henry II, when, according to the Pipe Rolls for the year 1169, the sum of £6 was expended on this bridge.

A grant of pontage was issued in 1372 for the bridge at Darlaston, and a bridge was also shown by Ogilby, but unfortunately without any note as to whether it was built of wood or stone. The present bridge is of no archæological interest.

At Walton, however, the road from Stone to Eccleshall crosses the Trent by a mediæval stone bridge having eight arches. Three of these are pointed in shape, one having two and another three ribs. They possibly date from the fourteenth century, but unfortunately there is no documentary evidence. The other arches are semicircular, built at a much later date, and later still the whole bridge was widened by about 2 feet on each side, giving a width between parapets of slightly over 16 feet. The bridge of Walton, for whose repair Roger Mareschal, *Walton Bridge*

ANCIENT BRIDGES

Aston Bridge

Canon of Lichfield, left in his will of 1317 the sum of £5, may possibly have been on this site.

Roger Mareschal also left the sum of £5 for the repair of the bridge at Aston, which crossed the Trent nearly two miles below Walton Bridge, but the present one is not ancient. The Trent and Mersey Canal runs close alongside the river at this point, and the present bridge is of the same type as the one over the canal.

Sandon Bridge

Sandon Bridge (Fig. 1) also has ribbed arches, one having four and the other five ribs. They are segmental in shape, and were probably built a century later than Walton Bridge. The total span is 20 yards, the width between parapets about 12 feet, and over each of the massive cut-waters is a recess to shelter foot-passengers.

A bridge is shown at Weston by Ogilby in 1674, but the present structure was built at least a century later. It has one segmental arch and semicircular string-courses at the road level, a design often found in this part of Staffordshire. Haywood Bridge, on the road from Great Haywood to Tixall, is yet more modern.

Essex Bridge

Chetwyn, writing in the year 1679, stated that Shutborough was 'formerly joined to Haywood by a wooden bridge, which being ruinous, was in ye last age rebuilt with stone and contains 43 arches, at ye end whereof stood ye Bishop's Palace.' Some of these arches possibly formed a causeway, for the present structure, now known as Essex Bridge (Figs. 2 and 3), has but fourteen arches over the river, spanning a total distance of 100 yards. Each cut-water is provided with a recess for foot-passengers, a very

1. Sandon Bridge.

2. ESSEX BRIDGE, GREAT HAYWOOD, FROM THE SOUTH-WEST.

3. ESSEX BRIDGE, GREAT HAYWOOD, FROM THE SOUTH-EAST.

necessary provision considering the length of the bridge and the fact that the width between parapets is only 4 feet.

In Leland's unpublished notes Wolseley Bridge is mentioned as 'Worsley or Worseley Bridge,' but Ogilby, more than a century later, shows it as 'Woolseley vulgo Ousley.' It was one of the most important crossings over the river Trent, and grants of pontage for its repair were issued in the years 1380 and 1387, also again in 1430. According to Shaw, Wolseley Bridge was 'blown-up' by the flood of 1795, and the present structure, which has three segmental arches, may be its immediate successor. Colton Mill Bridge on the road from Rugeley to Colton is certainly not more ancient. *Wolseley Bridge*

The road from Lichfield to Uttoxeter now crosses the Trent by an iron bridge, called High Bridge, erected in 1833. Shaw in 1798 describes the bridge of his day as 'an old erection of stone' 74 yards long, 11 feet wide between 'its low parapets,' and with seven pointed arches. He also adds that it was 'Newe builded 1622-3,' and widened in 1784. The original bridge, therefore, must have been extremely narrow. *High Bridge*

The bridge of 'Yoxhall' was mentioned in the Perambulation of Alrewas Hay (part of the Forest of Cannock), made in the year 1300, and the Patent Rolls for 1549 record 'an exoneration' of twenty pence for the maintenance of this bridge. In 1717 Sampson Erdeswicke, describing the village of King's Bromley, remarked that there had once been a bridge 'here over the Trent, but it is now decayed.' This may possibly have been 'King's Bridge,' which figured many times in the Records of the Stafford- *Yoxall Bridge* shire Quarter Sessions early in the eighteenth century. The repairs of Yoxall Bridge cost about £200 between the years 1731 and 1754, one-third of which was paid by the inhabitants of Yoxall. At the present time Yoxall Bridge has three segmental arches, with keystones which slightly project, probably dating from the middle of the eighteenth century. It spans a width of 31 yards and is 11 feet wide between the parapets.

Wichnor Bridges

Six miles to the north-east of Lichfield the ancient road called Rykneld Street crosses the river Trent by a series of arches known as Wichnor Bridges. The present structure is comparatively modern, but Wichnor Bridges were often mentioned in mediæval records, owing to the importance of this crossing. The Annals of Burton mention them when referring to a great storm in the valley of the Trent in the year 1255, while the Liberate Rolls for 1251 contained instructions to the Sheriff of Stafford to expend 10 marks upon 'the building of the bridge of Wicchenore.'

Shaw, in his *History of Staffordshire*, quotes an extract from a Tenure Roll of the reign of Henry III, which stated that 'in the Hay of Alrewas there were six oaks fallen, of which the king gave four to the bridge of Wychenour.' This bridge was also one of the boundary points of the Hay.

Tolls were evidently collected from persons using the bridge, as in the year 1308, when William de Saunford was indicted for harbouring a murderer, it was stated that he was the man 'to whom the King had granted certain customs for the repair of the bridge at Wychenore.' This grant is recorded in the Patent Rolls for the year 1307. There is no record to show if the mediæval bridge was built of timber or stone but it was evidently very narrow, as in the year 1760 Dr. Wilkes wrote that Wichnor Bridge 'will not admit any sort of wheel carriage, which is a misfortune to travellers.' In 1709 the sum of £25 was 'raysed' for its repair, a further £40 being required in 1755.

The flood of 1795 brought about a drastic change, and a new bridge was built, described by Shaw as 'a hansome bridge of three large arches, plain and neat in its architecture with the improvement of circular instead of pointed buttresses.' The present one has semicircular cut-waters, and is possibly the one described by Shaw.

The famous Trent Bridge at Burton was destroyed in 1864, when the present iron bridge was built. It had thirty-six arches according to Stephen Glover, but Sampson Erdeswicke (1717) and Henry A. Rye state that there were thirty-four, Rye adding that all but four were ribbed. The third arch from the western end fell during the flood of 1795 and was rebuilt the following year at the cost of £200. Leland gives no details, but states that there was 'a Chapel at the Bridge End,' the one doubtless for which Bishop Norbury in 1322 issued indulgences for forty days. It was dedicated to St. James.

Burton-upon-Trent

A bridge was evidently in existence at Burton as early as the twelfth century, as, according to Erdeswicke, William de la Warde in the year 1175 gave land for the benefit of the Bridge of Burton.

In February 1284 'protection' was granted, for a period of two years, to 'John Norff. monk of Burton upon Trent and keeper of the works of the bridge there, which has in great part been swept away by flood.' It safeguarded his men 'begging alms for the re-building of the said bridge.' A like 'protection' was given in December 1324, when the bridge was stated to be 'broken down.' Grants of pontage for its repair were issued in August 1383 and in May 1394, on each occasion for a period of three years. This bridge is the subject of an interesting paper read in January 1901 by Mr. H. A. Rye before the Burton-upon-Trent Archæological Society, which was published in vol. v. of their *Transactions*.

Swarkeston Bridge

Fortunately the greater part of Swarkeston Bridge has survived, although many of the arches have been reinforced with new vaults of blue brick, scarcely in keeping with the mediæval work. Several of the arches at the southern end are, however, unspoilt. They are pointed in shape (Fig. 4), and each arch has eight chamfered ribs.

The arches over the main stream date from the eighteenth century. Their form is semicircular, each with a string-course outside the arch-ring, and the cut-waters are also semicircular in shape. The span of these five arches is about 86 yards and the width between parapets is 22 feet. The mediæval arches have spans ranging from 11 to $18\frac{1}{2}$ feet in width. Ogilby, towards the end of the seventeenth century, showed a stone bridge of nine arches over the main river, and another with thirty stone arches a little farther south.

Swarkeston Bridge was the subject of an inquisition, taken in October 1275, at which it was stated that the merchants of the Soke of Melbourne had unjustly 'withheld the said passage money and tolls.' Pontage was granted in January 1325, December 1327, March 1338 (which covered both goods passing over and under the bridge), December 1347, and June 1355, the last being for the unusually long period of seven years. In the grant of 1327 it was called the bridge of Cordy, a name mentioned in the Derby Charter of 1204. A full list of the tolls for the grant of 1347 is given by the Rev. Charles Kerry in his article on 'Hermits, Fords and Bridge-Chapels' in vol. xiv. of the *Journal of the Derby Archæological Society*.

Repairs to Swarkestone Bridge cost the county the sum of £80 in the year 1682, and in 1713 it was ordered by the Sessions 'that chains be put across the arches owing to damage caused by boats and barges and none to pass under except by permission.' In 1745 a five-arch bridge over the main river was in existence; it was widened in 1802 at a cost of £1,780.

The chapel belonging to this bridge appears to have been built at its southern extremity, in a hamlet known as Stanton-by-Bridge. An inventory made in 1553 of the goods belonging to this chapel, published in vol. xi. of the Reliquary, recorded that 'We have a chappell edified and buylded uppon Trent in ye mydest of the greate streme anexed to Swerston bregge . . . and we saye that if the Chapell dekeye the brydge wyll not stonde.' Fortunately this prophecy failed, as the bridge still remains, although all signs of the chapel have gone.

Cavendish Bridge

The bridge, which carries the road from Derby to Loughborough across the Trent, was built, according to J. Nichols, by Sir Matthew Lambe and named Cavendish 'in compliment to the Devonshire family.' It took the place of a ferry, and when Nichols wrote his *History of Leicestershire*, at the end of the eighteenth century, a toll was still charged of sixpence for a chaise and tenpence for a wagon with four horses; further tolls being graduated 'for other things in proportion.' When the bridge was first opened the charges were the same as for the displaced ferry—*i.e.*, 2s. 6d. for a chaise and one penny for every person, whether on foot or horseback. This bridge (Fig. 5), which is still in use, has three large segmental arches across the river and a small landarch at each end. Each of the river arches has five ribs, a very unusual feature for a bridge built as late as the year 1758. The total span over the river is 48 yards and the width between parapets is 19 feet.

Harrington Bridge

Harrington Bridge, on the road from Nottingham to Ashby de la Zouch, took the place of Sawley Ferry. Leland records 'a stone Bridge with a Causey and many Arches partely over the very Gutte of Trent and partely for cumming to the Bridg by the Medoes for rysinges of the Trent.' This bridge could not then have been long built, as William Esyngwold and Richard Lister, on their ride from Nottingham to Gloucester in the year 1493, had to pay one penny for the use of 'Sallowe' Ferry. There is now an iron girder bridge at this site.

Bridges at Nottingham

In Nottingham where, according to the Anglo-Saxon Chronicles, there was a bridge as early as the year 924, the Trent is joined by a stream called the river Leen. The long series of bridges, which carried the roads from the south and east across the two rivers and the marshy ground between, were a continual source of trouble throughout their history, until they were replaced, in 1871, by the present structure. The part over the Trent was called 'Hethbethe Brigg,' but the name had many variations, and its origin is unknown. In the Close Rolls for 1252 there is the record of the gift of 10 marks towards the building of ' Hechebech ' Bridge, and grants of pontage for its repair were issued almost continuously, from the year 1311 to the middle of that century. The grant of 1314 was 'for the protection of Alice, late the wife of John le Palmer of Nottingham, who is building the bridge of Hethebeth.' In November 1323 and again in 1328 a commission was directed to audit her accounts for this bridge and another bridge 'newly built between it and Gameleston.'

In April 1335 it was reported that 'the great part is newly broken down,' and in March 1349 another commission was appointed to enquire who was responsible ' for the *stone* bridge over the Trent below Notyngham.' A further enquiry was ordered in January 1426, when the bridge over the 'water of Lene' was specially mentioned as being 'recently damaged by floods and divers carts.'

According to Thoroton, the repair of the bridge in the reign of King John was undertaken by the brethren of the Hospital of St. John in Nottingham, and certainly land ' at the bridge of Hebeye ' is mentioned in a grant to the Hospital made early in the thirteenth century. Archbishop Gray in 1231 granted indulgences to those contributing to the repair of the bridge of Hoybel at Nottingham, which is thought to be an alternative name for Hethbeth, and during the fourteenth and fifteenth centuries many wills included bequests towards its repair, the largest being £10 left by John Tawnesley in 1414 and the same sum by Robert Glade in 1424. Part of the bridge was evidently built of wood, for in 1324 Hugh Poyt was accused of stealing the 'timber of the bridge of Hethebethe.'

The bridge appears to have been completely out of use in 1363, as a grant of ferry was then issued, to be in force for a period of five years. It was therein specified that 'all the profits were to be employed upon the repairing and making of the bridge called Hethebethebrigg.' In 1397 Robert Bell was said to owe 100 shillings 'for the collection of the ferm of the bridge.'

The Records of the Borough of Nottingham contain many interesting facts concerning this bridge, including the accounts of the Wardens for several years. During a period of fifteen months, ending some time in the year 1458, the sum of £11 19s. 5d. was spent on repairs, and it is specifically stated that this 'excludes 24 great beams of the gift of Robert Strelly Esq. in the Park of Shipley, nor sixty loads of Basford Stone and five loads of poles, also gifts.' During the next three years a further £36 12s. 4½d. was expended.

The winters in those days were evidently very severe, as there are many references to the expense of clearing ice away from the bridge. An item for December 9th, 1486, reads: 'For brekyng of ise to ij men by the space of XV. dayes at morne and at evyn after ye grete frost—iiijs.' Another was for seven long poles 'for to make hokes and poyes' (levers) and 'a grete fellyng axe to hewe ise with—vjd.'

It seems probable that part of the bridge at that period consisted of timber beams laid across stone piers, as in 1486 two men were paid six shillings and tenpence for 'hewyng tymber at ye Brigges for ye arche next ye Chappell but oon,' and a further seven shillings was paid to 'William Rodes and his iij men for reryng of the seid arche and settyng up of hit by ye space of iij dayes and an half.' No mention is here made of masons' work, and the word 'reryng' would be more appropriate for the lifting of long beams than for the building of a masonry arch.

The name of the bridge changed from 'Heithbeith' to 'Trent Brigges' in the accounts for the year 1564-5, but that of the two small bridges over the pools, lying between the Leen and the Trent, remained as Cheney Bridges. They were only wooden structures even in the eighteenth century, and were so named from the fact that they were for use only in case of flood; at other times chains were placed across them.

Further trouble from ice was reported in the winter of 1621-2 and again two years later, while in 1638 the boat belonging to the Wardens was frozen in. A proposal was approved in 1643 to make a drawbridge over the Leen, and in 1684 George Wallis was appointed to take a weekly account of the expenses in 'the rebuilding of the Trent Bridges,' for which he was to receive the wage of 8s. a week.

In August 1698 instructions were issued to hold up all boats coming from Sawley and other places in Derbyshire, the arches to be 'chayned or stopped up' to prevent their passing. This no doubt was for the purpose of extracting toll, as a few years later orders were given for a toll-house to be built, it being specified that it must not cost more than £10 exclusive of the materials of a 'little old House in the Wood-yard there.' This was possibly the house 'at the East end of the bridge,' the building of which was mentioned in the accounts for 1511.

The rebuilding, in 1684, mentioned above was necessary owing to its destruction by a great flood in February 1683, when sheets of ice tore down a great part of the bridge. The frost lasted from September 1682 to February 5th, 1683.

The new bridge was built entirely of stone and, according to Blackner (1815), had seventeen arches, semicircular in shape, as shown in an engraving in his book. Charles Deering, who wrote in 1751, gives the number as twenty, and Sir Thomas Parkyns that of thirty-two, which included Leen Bridge. The bridge was severely damaged by floods in 1726, and in 1743 was said to be ruinous. Leen Bridge was 'shattered' by flood in 1795. Thoroton, in 1797, describes the Trent Bridge as 'an irregular link of arches, originally formed of rough stones, but now is disfigured with brick and the ordinariest materials.'

In the *History of the Old Trent Bridge*, by Mr. M. O. Tarbolton, the engineer responsible for the building of the new one, are reproductions of some excellent photographs of the bridge taken before its destruction, and from these it appears to have been a very massive structure. One arch at least was pointed in shape, with double arch-rings and ribs, and possibly dated from the fourteenth century, while the last arch at the north end, shown as almost touching the abutment of the new structure, is semicircular in form with five wide ribs springing from a plain impost moulding. The bridge had fifteen main arches, eleven with a span of 25 feet, the other four having an average of 20 feet. Its original width was about 12 feet, but the bridge had been widened on the upstream side with segmental arches. The foundations were formed of oak piles driven into the gravel bed of the river.

Like most mediæval bridges, Hethbeth Bridge had its chapel, founded by John le Paumer (Palmer) and Alice his wife in the year 1303, and dedicated to St. James the Apostle. Originally it had 'ij alters to Celebratte uppon,' but according to the Chantry Certificate Rolls of 1546 'there hathe ben butt oon Chaunteris preste there founde synes Anno xxvij (1535-6) and longe tyme before.' It was described as 'a Chappell covered with leade with an alter to Celebrate uppon the bridge of Nottingham aforesaid.' The 'Chapel of St. James the Apostle situated upon the Bridges over the Trent' was evidently served by the Prior of the House of Carmelite Brethren, as in July 1513 they sued for the sum of 2s. 8d. owed for celebrating mass for 'five weeks and a day.'

Below Nottingham there were until recent times no bridges across the Trent south of Newark, and the only means of crossing was by ferry. Ogilby's route from Nottingham to Lincoln crossed at Bleasley, but before reaching this ferry his road is shown crossing several bridges, two near Gedling, then Huntspil Bridge west of Gunthorp, a wood bridge over the

Derbeck River close to Hoveringham Ferry, and two miles farther on a 'wood bridge' over a 'Rill.'

Since its diversion, the main stream of the Trent passes more than a mile from Newark-upon-Trent, the river flowing through the town now being called the 'Devon.' It joins the Trent at Crankley Point, and the bridges across it will be described in a later chapter.

Kelham Bridge

This diversion was evidently made at the beginning of the thirteenth century, as in the year 1225 the tolls from the bridge of 'Kelum' were granted to the Burgesses of Retford for the sum of 20 marks of silver. The present bridge has five segmental arches, built of brick with stone arch-rings, and there are recesses at the road level, corbelled out from the side of the bridge. It is probable that all the earlier bridges on this site were built of timber, and the one in existence in the eighteenth century must have been a strange affair, being described by Throsby as 'a broad wooden bridge over the Trent of singular construction, apparently the most complex man ever formed.' Unfortunately he did not give an illustration.

According to Throsby, there was a tablet in the church of East Retford to the memory of Robert Sutton, who died in 1776, and amongst the many good works thereon recorded was the fact that 'he built Kelham Bridge'—presumably the one seen by Throsby.

In the *Memorials of Old Nottinghamshire*, by Mr. E. L. Guilford, will be found a most interesting map entitled a 'siege plan of Newark 1646.' From this it appears that both Kelham Bridge and the neighbouring one, leading to Muskham, had been de-

4. SWARKESTON BRIDGE.

5. CAVENDISH BRIDGE.

6. OKEOVER BRIDGE.

7. SANDY-BROOK BRIDGE.

stroyed. They are briefly dismissed with the remarks ' site of Kelham Bridge, now a bridge of boats,' and ' the place where Muscombe Bridge stood.'

Muskham Bridge, recently rebuilt in concrete, carries the North Road across the river Trent. The Patent Rolls for the year 1376 record a grant of pontage to ' the goodmen of Muskham by Newark' for the repair of Haybrigge and the causeway between Newark and that bridge. An earlier grant had been made in February 1358.

Below Muskham Bridge the river Trent even now is only crossed by two bridges, those at Dunham and Gainsborough, both comparatively modern. A bridge, however, formerly spanned the river between the village of Cromwell and South Collingham, but had been destroyed before the end of the eighteenth century. Dickenson records that remains of it could be seen in the year 1792, when the river was exceptionally low, and nearly a century later, when work was in progress to improve the navigation of the river, the piers of the bridge were found.

The ferry of Marnham is mentioned in the Patent Rolls for the year 1316, and Leland refers to the ' trejectus to Littleborough Village, whereby it is communely caullid Littleborough fery.'

According to Adam Stark, in his *History of Gainsborough*, the first stone of Gainsborough Bridge was laid in October 1787. It was completed four years later at a cost of about £10,000. The fine engraving in his book shows three elliptical arches, with the middle arch slightly larger than the other two. The ferry of ' Geynesburgh ' is mentioned in the Calendar of Inquisitions for the year 1336.

Gainsborough Bridge

CHAPTER II

THE NORTHERN TRIBUTARIES OF THE RIVER TRENT

THE river Trent, with its many tributaries, drains practically the whole of the North Midlands, embracing the counties of Stafford, Derby, Leicester, and the western half of Lincolnshire, but only those rivers which flow into the Trent from the north, together with their sub-tributaries, will be described in this chapter. Other tributaries will be dealt with in Chapter III.

THE RIVER BLITHE

This river is not of much interest as far as bridges are concerned, for most of them are of recent date. Girder bridges now cross the stream at Lower Leigh and at Upper Leigh, while Blythe Bridge is of concrete. The remainder are brick bridges, with the exception of the one near Crasswell Station and another near Burndhurst Mill, on the road from Uttoxeter to Stafford, which are built of stone and may date from the end of the eighteenth century. Some of the brick bridges are possibly also of this date, as Shaw, in his *History of Stafford*, mentioned that Blythford Bridge was built in 1766. Unfortunately he does not state if it was then a brick or a stone bridge.

THE RIVER DOVE

Rising on the eastern slopes of Axe Edge, some three miles south-west of Buxton, the river Dove forms for a great part of its course the boundary between the counties of Derby and Stafford. The majority of its bridges are of one pattern, having semicircular stone arches and string-courses, at the road level, of the same form. They are exceptionally well built, and usually have a width between parapets of about 16 feet. None have inscribed dates, but it is probable that they were built in the last half of the eighteenth century. Both Crowdecote and Glutton Bridges over the Dove, and Longnor Bridge, Windy Arbour Bridge, and several others over the Manifold, are of this type, as also is Hartington Bridge. The records of the Derbyshire Quarter Sessions for the year 1708 contain an item of £20 as 'a free gift' towards 'making Crowdycote Bridge,' but as Hartington Bridge at that time was described as 'a stone bridge for horses,' this pattern of bridge, which might for convenience be called 'the Staffordshire Type,' was evidently built later on in the eighteenth century.

The Staffordshire Sessions in 1709 ordered that 'a gratuity be given towards converting a wood horse-bridge over the Dove,' between Astonfield and Hartington, 'into a stone bridge for horses.' It is difficult, however, to determine the exact site of this bridge.

Viator's Bridge crosses the river Dove in Milldale at the northern end of Dove Dale. Its two arches are segmental in shape and the bridge is built of very rough masonry. William Bray in his *Sketch of a Tour into Derbyshire and Yorks*, written in 1777, mentions Milldale as the place 'where there is a little public-house by a bridge, which leads towards

Viator's Bridge, Milldale

Alstonfield and the great copper-mine of the Duke of Devonshire, called Ecton Mine.'

A mile west of Thorpe, the road to Ham crosses two branches of the Dove. Both bridges have single arches, segmental in shape. The one nearest to Thorpe bears the inscription 'St. Mary's 1790.' This bridge is 13 feet wide between its parapets, which seems to have been the original width; the other bridge is about 1 foot wider, having been widened on the upstream side by about 4 feet.

Coldwall Bridge

Coldwall Bridge, across the Dove, carries the now disused road from Thorpe to Cheadle, a milestone near the bridge giving the distance from the latter as eleven miles. The main arch and the two small ones in the causeway leading to the bridge are segmental in shape. At one time there must have been considerable traffic along this road, as the bridge has been widened by about 9 feet on the downstream side and is now 18 feet in width between parapets. The repair of Coldwall Bridge by order of the Derbyshire Sessions is recorded in the year 1717. It was then only a wooden bridge, and nine years later the Staffordshire Sessions appointed three persons to consider the building of a stone bridge in its place.

Okeover Bridge (Fig. 6), on the road leading to Mapleton, was probably built during the nineteenth century, but Mapleton Bridge over the Bentley Brook may be the one the rebuilding of which was authorised by Staffordshire in 1715, when the sum of £80 was granted. A further £55 was allowed in 1718 'as a half part with Derbyshire.' It has one semicircular arch, but this has been considerably widened. A mile or so above this bridge the Bentley Brook is crossed, near Sandy-brook Hall, by another stone bridge (Fig. 7), having one wide segmental arch between two smaller ones, semicircular in shape. It possibly dates from the eighteenth century.

In Mayfield the river Dove is crossed by a mediæval bridge (Fig. 8) which carries the road to Ashbourne. All five arches are pointed in shape and have chamfered arch-rings. Their overall width was originally only 10 feet, but the entire bridge has been widened on the downstream side to give a roadway of 18 feet. The Records of the Derbyshire Sessions mention the payment in 1682 of £50 for the repair of 'p'te of Hanging Bridge,' the other part being under the charge of Staffordshire. In 1766 a proposal was made at the Derbyshire Sessions to widen the bridge, and in the same year Staffordshire authorised expenditure on this bridge of a sum 'not to exceed £90 5s.' *Hanging Bridge, Mayfield*

A mile below Mayfield the Dove is joined by the Henmore Brook, which near its source is called the Scow Brook. None of its bridges are of special interest, but two in Ashbourne possibly date from the eighteenth century. The case of the School House Bridge was considered by the Sessions in the year 1698, when it was decided that 'a new foot-bridge of Timber be built and Mr. William James to have to his own use stone already laid down and the materials of the old bridge.' *Bridges in Ashbourne*

Between Ellaston and Norbury the Dove is crossed by a fine stone bridge (Fig. 9) which bears the date 1777. The two segmental arches have keystones which slightly project beyond the arch-ring, a rather unusual feature in this district, though common in the South of England. The width between parapets *Ellaston Bridge* is 15 feet, and recesses are provided above the cutwaters.

No bridge existed across the Dove at Rocester in 1786 according to Burdett's map of Derbyshire. A bridge over the Churn was, however, shown, but the present one is scarcely as ancient. This river, which in its upper reaches is called the Churnet, is crossed by several 'Staffordshire Type' bridges, the only one of note being the bridge at Alton, which has a fine stone balustrade in place of the usual parapet. According to the Records of the Staffordshire Quarter Sessions, it was ordered in the year 1687 that 'a cart bridge' be made at Cheddleston, provided that the inhabitants gave security for its maintenance, and in 1707 the sum of £100 was raised for building 'a stone cart bridge' over the Churnet at Oakamore. This bridge was not completed until the year 1717, and the total cost amounted to nearly £300.

In the neighbourhood of Uttoxeter the Dove is joined by the river Tean. The road to Rocester now crosses it by a girder bridge, but Hey Bridge, in Lower Tean, probably dates from the end of the eighteenth century. It has two segmental arches built of narrow bricks with stone keystones, a distinct variety from the usual type. The bridge between Upper Tean and Draycott has one stone arch with projecting keystones.

Dove Bridge

Dove Bridge (Fig. 10), a mile north-east of Uttoxeter, on the road to Doveridge and Derby, is on or near the site of a very ancient crossing. In the time of the Domesday Book the place was called Dubridge and in the thirteenth century Douebrigg. Three and possibly four arches of the present bridge

8. Hanging Bridge, Mayfield.

9. Ellaston Bridge.

10. Dove Bridge, Doveridge.

appear to date from the fourteenth century, but the two middle ones have been rebuilt. The ancient arches are pointed in shape with chamfered arch-rings, and one has four ribs. The cut-waters are very massive and have recesses for foot-passengers. The total span is 63 yards and the width between parapets about 24 feet, the bridge having been widened up-stream by about 13 feet in 1915. On the parapet is a stone inscribed with the date 1691, recording possibly the date of the rebuilding of the middle arches, as between the years 1689 and 1691 the Staffordshire Quarter Sessions allowed £350 for work on the bridge. These arches are semicircular and have small projecting keystones.

In 1748 it was reported to the Sessions that 'a certain horse bridge over the Dove lying in Aston, called Aston Bridge, is in danger of falling,' and it was decided that 'a carriage bridge ought to be built.' Whether this was done is not recorded, and the present bridge, which has three elliptical arches, appears to be of nineteenth century design. *Aston Bridge*

The bridge at Tutbury evidently required repair in 1717, although the sum of £160 was expended by Derbyshire during the reign of Queen Anne and £200 was required fifty years earlier to make good damage caused by flood; Staffordshire also spent nearly £400 on repairs between the years 1690 and 1740. It was described by Shaw in 1795 as 'an excellent stone bridge,' and in the *History of Tutbury* by Sir Oswald Mosley, Bart., published in the year 1832, it is shown as having five segmental arches with pilasters over the cut-waters. According to him this bridge was built in 1817, at the cost of £8,000, *Tutbury Bridge* about 20 yards below the old one, which was afterwards destroyed. The old bridge had nine pointed arches, and fortunately he gives an illustration of it taken from contemporary engraving. In the reign of King Henry IV there appears to have been a timber bridge at Tutbury.

Monk's Bridge, Eggington

The fine mediæval bridge near Eggington, on the road from Derby to Burton-upon-Trent, called Monk's Bridge (Fig. 11), has happily been saved by the building of a by-pass bridge which, besides freeing it from heavy traffic, enables the passing traveller to see the arches of the ancient bridge. These are four in number and segmental in shape, three having chamfered ribs. The style is like that of Warkworth Bridge, in Northumberland, Eamont Bridge, on the borders of Cumberland and Westmorland, and Framwellgate Bridge in Durham, which leads to the conclusion that the present bridge was built early in the fifteenth century.

Mr. Henry A. Rye, in his interesting 'History of Monk's Bridge,' which is published in vol. iv. of the *Transactions of the Burton-upon-Trent Archæological Society*, assumes it to be the bridge built by John de Stretton, Prior of Burton, towards the end of the thirteenth century, but this is scarcely possible, as the wide segmental arch was not in use at that period. At an inquisition taken in the year 1256 it was decided, as recorded in the *Annals of Burton*, that 'no one was responsible' for the repair of the bridge of Eggington, so John de Stretton rebuilt the bridge at his own cost. In the inventory of church property for the parish of Eggington, made in 1553, it was stated that two bells had been sold in 1549 to pay

11. Monk's Bridge, Eggington.

12. YORKSHIRE BRIDGE.

13. FROGGATT BRIDGE.

the expenses of 'repayrynge of the Muncks brydge, wch is so farre in decay that the township is not able to amend the same.' One arch was rebuilt in the year 1695 and the whole bridge was widened eighty years later, and now has a width of 17 feet between the parapets. The total span is 59 yards.

THE RIVER DERWENT

The upper part of this river now forms the reservoir of the Derwent Valley Water Board. Several small bridges have thus been destroyed, and others will find a like fate when any extension of the reservoir takes place.

At Ashopton it is joined by the river Ashop, here crossed by a small bridge named Cock Bridge. The single arch, segmental in shape, with a span of 9 yards, was originally about 6 feet in overall width, suitable only for packhorses. It has, however, been widened to give a roadway of slightly under 10 feet. *Cock Bridge*

A comparatively modern bridge crosses the Derwent at Ashopton, but a mile and a half below there is a massive structure called Yorkshire Bridge (Fig. 12) with two segmental arches having a total span of 24 yards. The cut-waters are semicircular in shape and have string-courses both at road level and at the springings of the arches. Its general design leads to the conclusion that it was not built until the end of the eighteenth century. Up to the year 1695 Yorkshire Bridge was of timber, but in that year it was decided by the Quarter Sessions to build a stone bridge with 'three Turned Arches' for the sum of £130. *Yorkshire Bridge*

Two miles below Yorkshire Bridge the Derwent is joined by the river Noe. The present bridges over this stream and its tributary the Hope Water are not of any special interest, and Mytham Bridge, over the Derwent, near Bamford Station, was evidently rebuilt during the nineteenth century.

Leadmill Bridge

Leadmill Bridge, a mile south of Hathersage, has three stone arches widened on both sides. It was formerly called Hazelford Bridge, and in 1708 the Sessions allowed the sum of £20 as 'a free gift towards making a bridge over the Derwent at Hazelford.' This amount was found to be insufficient, and a year later was increased to a total of £50 2s. 6d. for 'a wooden bridge with two stone piers and stone abutments.' These, however, were washed away 'before the wood was laid over.' It was then decided to build a stone bridge with three arches, but when completed it was found 'insufficient to receive the water in a large flood,' and a further arch and causeway was added at each end, bringing the total cost up to £166 6s.

Grindleford Bridge

The present Grindleford Bridge probably dates from the eighteenth century. It has three semicircular arches, built in two orders. The total span is 28 yards and the width between parapets is now 17 feet, the bridge having been widened on the upstream side. The Bridge of 'Gryndulforde' is mentioned in a grant made in the year 1407, and in 1706-7 the sum of £500 was spent on Grindleford, Froggatt, and Yorkshire bridges 'to make good the damage caused by heavy floods.'

Froggatt Bridge

A 'gurgit (or weir) at ffoggote' was referred to in the Baslow Court Rolls for the year 1483, but no mention was made of any bridge. Apart from the inclusion of Froggatt Bridge in the county records for 1688-9 and the fact that a bridge is shown on a plan of Stoke Hall estate, dated 1630, no historical information has come to light regarding this attractive bridge. It is very unusual to see a pointed arch (Fig. 13) with so wide a span. The width between parapets is only 9 feet, while the total span of the two arches is 30 yards. The bridge now carries little traffic, as the road over New Bridge, less than a mile farther south, has easier gradients. According to Mr. G. H. B. Ward's valuable article in the 'Coming of Age Number' (1921-2) of the *Sheffield Clarion Rambler's Booklet*, New Bridge bears the inscription 'A.D. 1781.'

Calver Bridge

Calver was evidently a river crossing of importance in mediæval times, as it had a bridge as early as the reign of King Edward IV. The Baslow Court Rolls for 1483 record the fact that 'Elotte ploughed from the common to Calvore Bridge, which never was done before.' One wonders if this was a notable achievement or if he was merely trespassing on someone else's property. Unfortunately the ancient bridge has gone and the present one is of no interest. Its semicircular cut-waters indicate a date late in the eighteenth century.

Baslow Bridge

Baslow Bridge (Fig. 14) is another case in Derbyshire where an interesting bridge has been saved from destruction by the building of a by-pass bridge. Its three lofty arches, nearly semicircular in form, span a total distance of 33 yards, and each arch has six narrow ribs. The piers, which are unusually narrow, have cut-waters with recesses at road level.

The width between the parapets is 12½ feet, and at the eastern end there is a small building for a watchman or toll collector.

Even at the beginning of the sixteenth century restrictions were placed on traffic over Baslow Bridge. In the year 1500 the jury ordered that 'no one henceforth shall lead or carry any millstones over the bridge of Basselowe under pain of 6s. 8d. to the lord for every pair of millstones so carried.' In the following year Thomas Harrison incurred the penalty for so doing. The bridge was evidently in urgent need of attention in the year 1649, when it was reported to the Sessions that 'without speedy repair itt will bee an undoinge to the Countye.' During the reign of Queen Anne repairs to Baslow Bridge cost the sum of £40.

Bridges in Chatsworth

A mile below Baslow the Derwent reaches Chatsworth Park, through which it flows for a distance of nearly two miles. The bridge leading to Chatsworth House from Edensor was built in 1760, at the same time as the stables, according to William Bray, but Mr. J. H. B. Ward gives the date as 1773. 'One Arch Bridge' (Fig. 15), near the Beeley Lodge, appears to be considerably older, being of ribbed construction, but it is possible that the bridge was only built in the seventeenth century or even later. The masonry is not unlike that of Cavendish Bridge over the Trent. The single arch is very impressive, the span being 23 yards and the rise approximately 30 feet. The bridge has a roadway 12 feet in width.

Bridges at Rowsley

Derwent Bridge, leading from Great to Little Rowsley, has five semicircular ribbed arches. It was widened in 1926, and now has a width between

14. BASLOW BRIDGE.

15. ONE ARCH BRIDGE, BEELEY.

16. SHEEPWASH BRIDGE, ASHFORD.

17. BAKEWELL BRIDGE.

parapets of about 40 feet. In 1682 the Sessions authorised the expenditure of £9 on Rowsley Bridge, and during the reign of Queen Anne £50 was spent on its repairs.

Half a mile below Great Rowsley the Derwent is joined by the river Wye, which rises a short distance from Buxton. None of its bridges above Ashford are of special note, but in Ashford itself there are three bridges of interest. The first, leading to the Marble Works, has two semicircular arches and probably dates from the time when the Marble Works were established in the middle of the eighteenth century. Sheepwash Bridge (Fig. 16), with three low segmental arches, carries the road to Sheldon, while the main road to Bakewell crosses by a bridge having several ribbed arches. The part nearest to Bakewell has four semicircular arches, each with seven ribs under the original part. The bridge has been considerably widened, and now is 27 feet between parapets. On one of these is inscribed ' 1664. M. HYDE,' said to be in memory of a doctor who was thrown from his horse and drowned. In 1776 it was reported that Ashford Bridge was ' so extremely narrow that it is dangerous for Carriages to pass over the same, and that the Battlements are very frequently knocked off by Carriages.' *River Wye*

At the northern end of Bakewell the river Wye is crossed by an attractive foot-bridge, known as Holme Bridge, which leads to Holme Hall. There are five segmental arches and also two small ones of semicircular shape. Over the former the width between parapets is slightly less than 4 feet. The main road to Baslow is carried by a massive bridge (Fig. 17) *Bridges in Bakewell* having five pointed arches, each with nine ribs. Five of these belong to the original structure, the others form part of the early nineteenth-century widening, which gives a width between parapets of about 24 feet. The total span is 40 yards.

Haddon Hall Bridges

Near Haddon Hall two bridges cross the river Wye—a foot-bridge with two obtusely pointed arches, having a width between parapets slightly under 3 feet, and a road-bridge with three ribbed arches of segmental form. The latter is possibly the one referred to in the Stewards' Accounts for the year 1663, when £97 was paid 'for making the new stone Bridge at Haddon.' In the archives of Haddon Hall, published in volume xv. of the *Journal of the Derbyshire Archæological Society*, is the mention in the grant of the year 1505 of land 'near the bridge called "Coteyerte,"' and in an earlier grant 'of all his meadow ... lying in the meadow of Horscroft, between the meadow of the lord and the Stone bridge.'

River Lathkill

Less than a mile below Haddon Hall and a few hundred yards after passing under Fillyford Bridge (late eighteenth century) the Wye is joined by the river Lathkill. At Alport this stream is crossed by a stone bridge, bearing the date 1793, which has one segmental arch with projecting keystones and a roadway of 18 feet. This was not the first bridge to be built at this spot, for in the year 1718 the Sessions ordered a horse-bridge to be built at Alport Ford over the Lathkill. It was stated to be urgently needed, as 'great gangs of London Carrier's Horses, as well as great drifts of Malt Horses, and other dayly carriers and passengers,' used the ford.

18. Darley Bridge.

19. Cromford Bridge.

20. MERRIEL BRIDGE.

21. NEW BRIDGE, BLYTH.

Darley Bridge (Fig. 18), the first crossing over the *Darley* Derwent below Rowsley, still has two pointed *Bridge* arches, each with four ribs; the remaining three have been rebuilt with semicircular arches, the type also used for the widening of the bridge on the upstream side. It is possible that other arches exist in the approaches of the bridge, but hidden by accumulated soil, as in 1682 it was reported to the Sessions that of the 'Seaven Arches, three of the said Arches are filled up.' It was ordered that the fifth arch should be put into use by cutting a channel, 25 yards in length above and 40 yards below the bridge and about 8 yards in breadth, to give a free passage of the water. This was estimated to cost £40, but from the accounts, for the reign of Queen Anne, this was evidently exceeded by the sum of £6.

Matlock Bridge is another Derbyshire bridge with *Matlock* pointed arches. They are four in number with very *Bridge* massive cut-waters. Originally the bridge was only about 13 feet in overall width, but it has been widened upstream by about 18 feet; both parapets are comparatively modern.

Leland paid little attention to Derbyshire bridges, *Cromford* but he did mention Cromford Bridge. It was prob- *Bridge* ably the one that exists at the present time, as the four-centred arch can usually be assigned to the fifteenth century. Cromford Bridge (Fig. 19) is the only one of this type over the Derwent, and fortunately it has not been spoilt by the widening on the upstream side. The width between the parapets is now slightly over 20 feet, but that of the original bridge was only about 12 feet. The total span of the three arches is 32 yards. On a stone is inscribed

THE LEAP OF MR. B. H. MARE JUNE 1697,' but the story of this feat is not recorded.

Whatstandwell Bridge

Among the muniments belonging to the Duke of Rutland is an agreement made in the year 1390, between Thomas, Abbot of Derley, and John de Stepul, reciting that 'John intends, for the weal of his soul, to construct a bridge at his own cost over the Derwent next the house which Walter Stonewell had held of the Convent, where no bridge had ever been constructed, of which bridge the east end must stand on their land, and the abbot and convent grant that it may so stand for 24 feet.' This appears to be the history of the bridge now known by the quaint name of Whatstandwell Bridge. Leland gives the name as Watstonde Wel Bridge, although in an agreement of sale made in the year 1510 the place was known at Watstanwell, the bridge being one of the specified boundaries. According to Stephen Glover, 'Watstandwell Bridge' was rebuilt in the year 1795 with seven arches, and the present one, which has seven semicircular arches, is probably the bridge to which he referred.

River Amber

At Ambergate the Derwent receives the river Amber, which rises a few miles north-west of Claycross. It is crossed by numerous bridges, most of which are quite modern. Titus Wheatcroft, in a memorandum made in 1722, published in vol. xix. of the *Journal of the Derbyshire Archæological Society*, gives a list of seven 'good stone bridges' which existed in his day. He gives their names, but no further particulars.

Belper Bridge

About two and a half miles south of Ambergate the river Derwent reaches Belper, where it is crossed by another stone bridge, probably the actual one completed in 1798 at a cost of £2,220; the earlier bridge, for the repairs of which the county paid the sum of £71 early in that century, having been washed away.

At Duffield, until early in the nineteenth century, there was only a narrow bridge across the Derwent, and most vehicles used the ford. In 1658 the inhabitants of Duffield were presented for 'not repaying and mending a Foarde in ye river of Derwent,' and in 1714 the sum of £30 was voted by the Sessions for the repair of the ford, it being stated that 'the same being of great use to preserve Duffield Bridge.' This bridge, widened in 1803 at the cost of £1,025, still stands. It has three segmental arches spanning a distance of about 100 yards. The arch-rings of the western arch are built in two orders, both having moulded edges. The other arches appear to be of later date. At the present time the width between parapets is 25 feet, about double that of the original structure. *Duffield Bridge*

In the fifteenth century it may be that the bridge was of timber, as in the year 1411 the Master Forester of Duffield Frith was ordered to deliver to the tenants of Duffield 'sufficient wood for the rebuilding of their bridge over the Derwent.' The Churchwardens' Accounts for Duffield are published in vol. xxxix. of the *Journal of the Derbyshire Archæological Society*, and although they make no mention of expenditure on the bridge, the following item for the year 1781 is of interest: 'An Umberrello for the use of Minister Mr. Ward—13s. 6d.'

At one time a bridge crossed the Derwent at Little Chester, on the northern outskirts of Derby, but it *Bridges in Derby* had been destroyed before the middle of the seventeenth century. The foundations were seen by Stukely and also by Camden, while Hutton, in 1791, said that he could feel them with an oar and had seen them when the water was clear. It is a pity that no one examined them in the year 1661, when, according to Hutton, the Derwent was dried up and 'people walked over the bed of the river dry shod.'

The 'Plan of Darbye in 1610' in Robert Simpson's *History of Derby* (1826) shows only one bridge across the Derwent, but his later map gives three—'St. Mary's,' 'Exeter,' and another near the Rolling Mills. The Patent Rolls for September 1325 record a grant of pontage for a period of three years, which was renewed in 1328, and a fresh grant was made in March 1329, the reason being that 'the collectors first appointed have shown themselves trustworthy.'

St. Mary's Bridge had a chapel, and in the year 1360 Bishop Robert de Stretton issued a licence to the Guardians of the Bridge of Derby 'for celibration in the Chapel on the Bridge for two years.' This building, which has recently been repaired as a memorial to the Haslam family, was used in the seventeenth century as a factory, as in July 1640 the Mayor and Burgesses made a lease with Hamlet Barne, Clothier, for 'all that messuage, burgage or tenement called St. Mary Chappell . . . to keepe fortie poore inhabitants of this Borough in work.' According to Glover, the chapel was used by the Presbyterians during the reign of Charles II, and was later inhabited by Thomas Eaton, a surgeon, while Hutton wrote: 'It forms part of the bridge with which it is interwoven as if erected with it, and was in my time converted into little dwellings.'

Hutton also remarked that St. Mary's Bridge was very lofty and narrow, and that 'An Act and £4,000 are procured to erase that nuisance . . . and to erect another bridge.' This plan was evidently carried out, for Glover states that St. Mary's Bridge was rebuilt in the year 1788. It has three arches, semicircular in shape, as are also the cut-waters. The total span is 50 yards, and the width between the balustrades is 30 feet. Remains of part of the former bridge can still be seen below the chapel. A very full description of this bridge and its chapel, together with its history, will be found in vol. v. (new series) of the *Journal of the Derbyshire Archæological Society*.

The Markeaton Brook flows through the town of Derby, and is crossed by many bridges. None are ancient, and Hutton at the end of the eighteenth century remarked that there were ten bridges over this stream, 'four paltry ones of stones, and six more paltry of timber; none of them is passable in a flood.'

Exeter Bridge, which has been recently rebuilt, is a successor to one erected early in the nineteenth century as a private bridge and opened to the public about 1810.

THE RIVER IDLE

Three rivers—the Poulter, the Medan, and the Maun—join together near Elkesley to form the river Idle. The first of these rises on the borders of Derbyshire and flows through Clumber Park, receiving on the way the outflow of the Great Lake of Welbeck Park. The river Medan runs almost parallel to the Poulter, about three miles to the south-east, through Church Warsop and Thoresby Park, while the Maun flows through Mansfield, Clipstone, Edwinstowe, and Ollerton. Most of the bridges are comparatively modern, but some may date from the end of the eighteenth century. Ogilby showed a wooden bridge in Mansfield on the road to Nottingham, and according to Throsby the bridge at Ollerton was 'thrown down' by flood in 1795. Evidently even its successor has gone, as the river is now crossed by a concrete bridge.

Merriel Bridge

Between West Drayton and Old Eel Pie House the ancient road to Bawtry crosses the river Maun by a small stone bridge (Fig. 20) which has two pointed arches, with chamfered arch-rings, the only bridge of this type in the county of Nottingham. Its total span is only 9 yards, but the width is now about 18 feet, the bridge having been widened by means of concrete girders.

The 6-inch Ordnance map of 1900 shows a narrow bridge on this site, but gives no name to it. The map of 1676 in Thoroton's *Antiquities of Nottingham* shows 'Mirihilbridge' near this place, which is probably the correct name for this bridge.

Thoroton also states that Merriel Bridge 'lies at the entrance to this township (Merrielbridge) on the Yorke Rode Way betwixt Turford (Tuxford) and Scroby.' This is the road shown by John Armstrong in 1776 as the 'Forest Road,' branching off from the present North Road at Markham Moor and rejoining at Barnby Moor. The construction of the North Road through East Retford was authorised by Act of Parliament passed in 1766, and the preamble of the Act stated that 'the road from Little Drayton to a certain bridge, called Twyford Bridge, is narrow and ruinous.' This evidently referred to the 'Forest Road.'

The present bridge at East Retford may be the one 'partly rebuilt and widened' in 1794. There were, however, earlier ones, as Leland records a bridge but gives no details, and in the year 1385 pontage was granted to 'the bailiffs of Retford-in-the-Clay for the repair of the bridge over the Idel.' *East Retford Bridge*

Leland mentions a stone bridge at Mattersay, but the present one is of late eighteenth-century design. The three stone arches are semicircular in shape with brick string-courses, and there are stone pilasters over the cut-waters. The width between the parapets is slightly over 10 feet. In the seventeenth century the tolls of Mattersay Bridge belonged to the Corporation of Nottingham. *Mattersay Bridge*

The ferry of 'Bautre across the Iddel' is mentioned in the Calendar of Inquisition for the year 1343, and there does not appear to be any record as to when the first bridge was built. The present one, built probably during the nineteenth century, has three segmental arches spanning a distance of nearly 30 yards. The roadway is 20 feet wide. *Bawtry Bridge*

A short distance above Bawtry Bridge the river Ryton joins the Idle. This stream rises a few miles west of Worksop and acts for the first half of its course as a feeder to the Chesterfield Canal, which was begun in the year 1771. It is probable that *River Ryton* several of its bridges date from this period, but the only one of importance is New Bridge, half a mile west of Blyth, on the road to Rotherham. This bridge (Fig. 21) has three semicircular arches and a handsome stone balustrade.

CHAPTER III

THE SOUTHERN TRIBUTARIES OF THE TRENT

THE RIVER SOW

THIS river rises seven miles west of Eccleshall, close to the borders of Shropshire and Staffordshire. Between Chebsey and Great Bridgeford it is joined by the Meece Brook and two miles below Stafford by the river Penk, which drains the district between Wolverhampton and Stafford.

The Sow is crossed by numerous bridges, but none of these are ancient, and the same may be said concerning its tributaries, and unfortunately nothing is known of the history of those which preceded them. Stableford Bridge over the Meece Brook was shown by Ogilby on his route from Stone to Nantwich, and the Liberate Rolls of the reign of King Henry III contain instructions to the Sheriff of Stafford 'to cause the Bridge of the King's Fishpond at Stafford to be repaired whenever it requires reparacion.'

THE RIVER TAME

The river Tame rises near Walsall, and in the Calendar of Deeds belonging to that town is the record of an enquiry held in the year 1606 regarding the liability ' to keep in repair a bridge called Tamebridge and a wooden bridge and way adjoining as far as a place called Fryars park corner.' It is difficult to identify this actual bridge, but possibly it is the one on the road from Walsall to West Bromwich which was shown as Tame Bridge on William Yates' map of Staffordshire, published in 1775.

Another bridge over this river named by Yates is Perry Bridge, on the road from Birmingham to Bromhills, which has four semicircular arches. It is probably the one built by order of the Staffordshire Quarter Sessions, held in 1709, to take the place of a 'wood horse bridge.'

River Rea   About four miles below Perry Bridge the Tame is joined by the river Rea, which, rising in the Lickey Hills, flows through the eastern part of Birmingham. Leland, approaching Birmingham from the south by way of King's Norton, records that he came 'through a pretty Street or ever I entered Bermingham Towne.' He entered through the district called 'Dirty' by him, but now named Deritend. To reach the town itself it was necessary to cross the river Rea, and he goes on to say that 'I went through the Ford by the Bridge.' From this it seems that the bridge at that time was for foot-passengers only, and presumably he was on horseback. An engraving of Birmingham in 1732 by Westley shows a bridge with four segmental arches across the Rea between Digbeth and Deritend, having recesses over the cut-waters. A timber foot-bridge is also shown a short distance further downstream.

Water Orton Bridge   At Water Orton a massive stone bridge (Fig. 22), having six semicircular arches with chamfered archrings, crosses the river Tame. The total span is 45 yards and the width between the parapets is slightly over 10 feet. Over each cut-water is a recess for foot-

22. WATER ORTON BRIDGE.

23. COLESHILL BRIDGE.

24. HEMLINGFORD BRIDGE.

25. ELFORD BRIDGE.

passengers. John Harman, Bishop of Exeter, who held the manor of Sutton Coldfield, built bridges at Water Orton and at Curdworth early in the sixteenth century. The latter has been replaced by one built of brick, but it is possible that the present bridge at Water Orton may be the one he made. It is unusual, however, to find a sixteenth-century bridge with round arches, but on the other hand there appears to be no record of the rebuilding of this bridge. A former bridge was the subject of an indulgence, for forty days, granted by Reginald Bowlers in the year 1459.

About two miles below Curdworth Bridge the Tame is joined by three rivers—the Cole, the Blythe, and the Bourne. Most of the bridges over these streams have been recently rebuilt or extensively widened, but each river has one bridge of interest. *River Cole*

At Cole End, the northern part of Coleshill, there is a very massive bridge (Fig. 23) with six segmental stone arches. These are built in two orders, both being chamfered. Unfortunately the history of this bridge is unknown, but it is probably considerably older than Water Orton Bridge. The total span is 41 yards and the width between parapets is 13 feet. Originally it was scarcely more than half this width, but the upstream side has been widened with brick. *Coleshill Bridge*

The one notable bridge on the river Blythe is a narrow packhorse bridge alongside the dangerous ford on the track leading from Hampton in Arden to Marsh Farm and Berkswell. Three of the arches are pointed, the other two being segmental. There are massive pointed cut-waters upstream, but those on the downstream side have square ends, and on *Hampton in Arden Bridge* one of the piers is the base of a cross. An excellent scale drawing of this bridge is given by J. A. Cossins in his paper in vol. xlii. of the *Transactions of the Birmingham Archæological Society.*

Furnace End Bridge

The Bourne, much the smallest of these three rivers, rises near the village of Fillongley. Between Furnace End and Shustoke Station it is crossed by a mediæval bridge with one semicircular arch spanning about 12 feet. This arch, which has four chamfered ribs, was formerly about 12 feet wide, but was widened in 1924 to nearly 40 feet. Mr. Cossins also gives a sketch of this bridge in which the ribs of the ancient arch are clearly seen.

Hemlingford Bridge

At Hemlingford, three miles north-east of Curdworth, a typical late eighteenth-century bridge (Fig. 24) crosses the river Tame. The five segmental stone arches have a total span of 60 yards. Two cut-waters on each side of the bridge have small triangular recesses at the road level, while over the remainder are stone pilasters. The distance between the parapets is 14 feet, and on one of them is carved a number of names, including—Richd. Worthington, New House; William Harrison, Drakenidge; William Brown, Starley; William Bond, Kingsbury Hall; also Richard Astley and Henry Cooper.

At Fazeley, two miles south of Tamworth, Watling Street crosses the Tame. It was evidently considered an important crossing in mediæval times, as the Patent Rolls for March 1313 record the appointment of a commission 'to view the bridge across the river Thame between Stretford by Wilmundecote and Faresleie and to compel those who are liable to execute all necessary repairs.' Leland 'passed over Faseley Bridge of 16 arches of Stone over Tame,' but Shaw at the end of the eighteenth century gives the number as thirteen. Little can now be seen of this ancient causeway.

In Tamworth the Tame is joined by the river Anker which, with its tributary the river Sence, drains the vale lying north-east of the road from Nuneaton to Tamworth. Harris Bridge across the Sence may date from the end of the eighteenth century, but Fielden Bridge over the Anker, a mile north of Atherstone, is of later date and is certainly not the stone bridge with eight arches seen by Ogilby. In March 1332 the Abbot of Merevale was granted pontage for a term of three years for the repair of 'Feldenbrigg by Atherstone over the river Ancre.' *River Anker*

The bridge in Polesworth has ten semicircular arches and bears an inscription 'H.E. 1776. T.S.' The arch-rings are of stone, but the rest of the original part of the bridge is built of brick. In 1924 the bridge was widened in concrete to give a width of 40 feet between the parapets. Bole Bridge on the eastern borders of Tamworth was the subject of a grant of pontage in the year 1442; Leland gives the name as 'Bowebridge.' *Polesworth Bridge*

Close to Tamworth Castle the road to Coleshill crosses the river Tame at Lady Bridge. Leland calls it St. Mary Bridge, 'havinge 12 great Arches,' and Shaw records that the bridge was destroyed by the flood of 1796. The present bridge, which has six semicircular arches, is probably the one that took its place. The total span is nearly 90 yards and the width between parapets is 23 feet. *Lady Bridge*

Hopwas Bridge

The road from Tamworth to Lichfield crosses the Tame in the village of Hopwas, where Leland 'passed over a Stone Bridge of . . . Arches bearinge the Name of the Village.' Unfortunately he forgot to state the number of the arches, which according to Shaw numbered sixteen at the end of the eighteenth century. It also had been damaged by the great flood. The bridge now has six elliptical arches.

Elford Bridge

At Elford, about four miles north of Tamworth, two branches of the river Tame are crossed by a bridge built of brick and red sandstone. The eastern part has two semicircular arches with brick vaults, the remainder being of stone. The western section (Fig. 25) has eight arches, of which two are segmental and the rest semicircular. These also have brick vaults, except for one, which is constructed throughout of sandstone. Both bridges have a width of 16 feet between parapets, and were probably built during the eighteenth century. Shaw describes it as a 'hansome stone bridge.'

Chetwynd or Salter's Bridge

A mile from its junction with the Trent the river Tame reaches a bridge known as Chetwynd or Salter's Bridge. Leland calls it by the latter name, but gives no particulars of its construction. In 1589 it was described as a stone bridge 'in great decay,' and twelve years later, according to the Alrewas Register Book, Salter's Bridge, being 'greatly in decaye and broken downe, was new begonne and made much broder by two foote' at a cost of £200. Between the years 1690 and 1756 more than £100 was spent by Staffordshire on repairs to Salter's Bridge. The present one has three cast-iron spans.

26. ENDERBY MILL BRIDGE.

27. OLD BRIDGE, BELGRAVE.

28. REARSBY BRIDGE.

THE RIVER SOAR

John Nichols in his *History of the County of Leicester*, published early in the nineteenth century, mentions several bridges across the river Soar near its source in the extreme south of the county. According to him, Lion's Bridge between Frolesworth and Claybrooke had been 'entirely destroyed and not replaced,' and Stoney Bridge 'on Foss-road' was 'a curious arch of forest-stone' (granite).

During the seventeenth century the road from Sharnford to Leicester must often have been impassable, as Ogilby in his *Road Book* made a special note that a mile north of Sharnford 'The River Soar Flows in ye Road at Winter Floods.' He showed Langham Bridge as a 'Stone Br. 1 Arc over a Brook,' but did not actually give the name. Langham Bridge had five arches, with the warning, 'Winter floods.'

Throsby in 1797 remarked that 'Langham Bridges extend like a chain along the Fosse ... most ancient work of labour ... seemingly without design, but that of durability.... without a fence-wall on either side.' This bridge evidently lasted until late in the nineteenth century, as in 1892 it was described by Colonel Bellairs as a mediæval bridge with pointed arches and recesses on the piers, widened on one side by small brick arches. The present brick bridge contains the remains of an older one of stone. Nichols records that at Narborough about the year 1788 there was built 'by subscription a bridge over the river Soar.' This probably gives the date of the brick bridge near Narborough Station on the road to Little Thorpe.

Enderby Mill Bridge

Just over a mile below Narborough there still exists near Enderby Mill the remains of a fine mediæval bridge (Fig. 26) on the line of the old road from Blaby to Enderby. It has two stone arches, pointed in shape, with chamfered arch-rings. The total span is 12 yards, the overall width 6 feet. There are no parapets, but on each side is a massive cut-water, and the bridge stands quite isolated in a field and can be seen from the modern road.

Less than half a mile below Enderby Mill the Soar is joined by the river Sence. None of its bridges however, have archæological interest, but three miles farther north, at Aylestone, the river is crossed by a narrow foot-bridge nearly 50 yards long, with several pointed arches and very massive cut-waters. The width between the parapets is slightly over 4 feet.

Bridges in Leicester

In mediæval times three bridges crossed the river Soar in Leicester—the North Bridge, known later as St. Sunday's Bridge, Bow Bridge over a branch of the river, and the West Bridge on the road to Hinckley. At an inquisition taken in the year 1253 to deal with the question of tolls called 'Brigg-silver,' it was stated that at the end of the eleventh century Earl Robert gave a piece of land near the bridge to one Penbrioch to build on 'for his more convenient gathering of these penies.' The tolls, amounting to a penny for six cart-loads and one halfpenny a horse-load a week, were charged on all dead wood and 'boughs blown down by the wind' fetched from the Forest of Leicester. The toll on 'one man's load a week' was only one farthing. Land 'by the North Bridge' was also given by him to the 'church of our

Lady.' The payment of Brigg-silver was remitted by Simon de Montford about the year 1254 in return for an annual payment. The first of the Merchants' Guild Rolls (*c.* 1225) records various payments made for work done on the North Bridge, as does the second Roll of 1262.

In the year 1300 the sum of £21 3s. 2d. was expended on the bridges. The North Bridge was then evidently a stone one, as the accounts for 1307 include an item—'the wage of a certain mason for mending an arch of North Bridge and two workmen helping him, 2s.' Another bridge described as being 'near the house of John the Painter,' possibly Little North Bridge over a small stream, was, however, of timber, as in 1316 there was an item of 'Wood for planks, in sawing and carpentry, 2s. 4d.' Eleven years earlier John the Painter was charged with carrying away '3 pieces of wood from (Little) North Bridge.' He was told 'to bring back these pieces,' but no other penalty was recorded. This bridge is now known as Frogmire Bridge.

The West Bridge was also of stone, for, in the year 1307, the sum of £2 3s. 7½d. was paid by the Keeper of the West Bridge for work, including labour, lime, stone, etc., and in the Mayor's Account for 1365 appears an item, 'Three Masons at 3s. 4d. a week each,' and also two shillings for a centre (sceynture) for supporting the arch. This arch had evidently collapsed, as a further item was—'Wages of divers people for collecting stones there in the water.' The total cost was £4 8s. 7d.

In December 1330 pontage was granted for a period of three years on 'wares passing over the bridges of the town of Leicester for the repair of the same bridges.'

'Sent Sunday Brigg' appeared in the *Book of Acts* of 1550, and after this date the name of 'North Bridge' was only occasionally used, but there is no record of any reason for the change. As at Nottingham, ice was a menace, and the Chamberlain's Accounts for the year 1683 include the payment of 2s. 'for Ale for severall persons that helped the secureing of the West Bridge upon the Thawe of the great frost from the Ice that came down the Water.'

Bickerstaffe described St. Sunday's Bridge as having 'eight wet arches, the midmost high and wide; two more on the town side, small and useless . . . 98 yards 1 foot long . . . 5 yards 2 feet wide'; and added that 'one of its arches nearest the town is pointed; the other nine are round.' This was evidently the bridge destroyed by flood in 1795, as the one shown by Nichols has segmental arches with narrow piers and low cut-waters. Leland gives the number of arches as 'seven or eight.'

West Bridge, according to Nichols, had four arches and a dwelling-house 'once a chapel where two mendicant friars asked alms for the benefit of the priory.' A picture of West Bridge, showing the chapel, is given by Mr. C. J. Billson in his book entitled *Mediæval Leicester*.

Bow Bridge, carrying a path to Glenfield from St. Augustines, was shown on a map of 1642, in Thompson's *History of Leicester*, as 'ye Bowe Bridge.' It had three arches, and close by, to the north, was 'Lesser Bowe Bridge.' A picture of these bridges given by Nichols shows one with a single segmental arch, having double arch-rings and a third ring at the crown, while in the background can be seen three semicircular arches of the other bridge, which has narrow piers with recesses over the cut-waters. Another engraving shows the second bridge as having four pointed arches. In 1666 the sum of £15 12s. od. was paid for the repairing of Bow Bridge, but in 1791 it was washed away.

According to the Leicester Chronicle for 1862, 'Bow Bridge, recently removed, had five semicircular arches six feet wide with nitches at intervals.' Stakes and faggots were found below the piers, also a skeleton.

Leland, approaching Belgrave from the west through Bradgate, 'cam over a great Stone Bridge' which Ogilby, more than a century later, showed as having seven arches. Part of the present bridge (Fig. 27) is ancient, but it has been widened on both sides and the roadway is now 21 feet wide. It has seven arches which are masked by the widening except for one pier that can still be seen, as, on one side of the bridge, a single arch has been used to span two of the original arches. The ancient part is scarcely 8 feet in overall width. *Old Bridge, Belgrave*

At an inquisition taken in the year 1357 mention is made of Belgrave Bridge 'for horses, carts and carriages.' The complaint was that Roger de Shepeye and eight other men from Belgrave had 'set bars, fixed across piles and fastened with locks and keys, across the said bridge.' They were ordered to remove the obstruction at their own expense.

About five miles north of Leicester the Soar is joined by the river Wreak. The upper part of this *River Wreak* river is called the Eye, but below Melton Mowbray the name Wreak is used.

Prior's map of Leicestershire, published in the year 1779, shows no road from Stapleford to Wyfordby, but, according to Nichols, the Earl of Harborough in 1773 built a bridge, with a single span of 32 feet, between these two villages. It took the place of a foot-bridge said to have been built by seven brothers and seven sisters of the house of Sherard (the family name of the Earl of Harborough), each brother building an arch, the sisters completing 'the other part of the bridge.' This doubtless referred to the financial aspect of the case, and not to the actual building of the bridge.

None of the existing bridges in this district call for remark, but numerous references are made to Burton and Kettleby Bridges in the Town Accounts of Melton Mowbray for the seventeenth century, published in the *Transactions of the Leicestershire Architectural and Archæological Society*.

Nichols, writing about Melton Mowbray at the end of the eighteenth century, recorded that 'There are two bridges over the Eye on the South and West, one with six, the other with seven large pointed arches; on that leading to Okeham is an inscription: "This bridge was repared Anno Dom. 1659, again in 1785."' Also that on an arch of the other bridge was carved: 'This arch was erected at the expense of the Right Hon. Robert, Earl of Harborough 1775.'

The bridge at Ashfordby, however, which crosses the Wreak on the road to Kirby Bellars, was formerly a packhorse bridge about 7 feet in overall width. It has three stone arches, semicircular in shape, but the widening has been carried out in brick.

Another bridge of this type, between Queniborough and Rearsby, carries the main road from Leicester to Melton Mowbray over a tributary of the Wreak. It has fortunately been widened on the downstream side only, so that the rough masonry of the ancient bridge is still visible. It is very difficult to state the age of this style of bridge, built evidently of local materials and by very unskilled labour.

At Rearsby, close to the church, there fortunately still remains a packhorse bridge (Fig. 28) in an almost unspoilt state. It has seven semicircular arches spanning a distance of nearly 20 yards and is 5 feet in width between the parapets.

A bridge evidently existed at Thrussington early in the fourteenth century, as mention was made at an inquisition of 1326, dealing with Thrussington, of land ' at the head of the bridge.' The present one is not ancient, neither is Lewin Bridge which carries the Fosse Way. During the eighteenth century the Wreak was made navigable to Melton Mowbray, and at the same time Lewin Bridge was ' repaired and enlarged.' According to Nichols, it ' had long stood decrepid on its worn out stumps.'

Nichols also refers to a bridge with three arches across a brook at the entrance to Syston, called ' The Nine Days' Wonder,' having been built by three bricklayers and six labourers in the space of nine days. During this short time they laid 25,000 bricks in addition to 150 tons of stone. It certainly well deserved its name.

THE RIVER ROTHLEY

Half a mile or so after receiving the river Wreak the Soar is joined by another stream, the river Rothley. At Anstey, about three and a half miles from the middle of Leicester, on the continuation of an old trackway called Anstey Lane, is another packhorse bridge (Fig. 29) with five arches spanning a distance of 18 yards. The parapets are 5 feet apart, and there is a recess for foot-passengers over each cut-water. It is a particularly fine specimen of this type of bridge, but unfortunately its history appears to be unknown. Half a mile below is a small bridge of the same type leading from Anstey to Astill Lodge which, according to the 6-inch Ordnance map, is called King William's Bridge. If the name referred to William III, it would indicate that these bridges were built late in the seventeenth century.

Another narrow bridge called Sandham Bridge crosses the Rothley behind Bishop Latimer's house in the village of Thurcaston.

The road from Rothley to Cossington must have been of considerable importance in the Middle Ages, as in May 1331 pontage was granted for a period of four years for the bridge across the Soar between 'Cosyngton and Rothele.'

From Cossington the river Soar flows past Mountsorrel under an iron bridge, which replaced one of stone, to Barrow-upon-Soar. Here it is crossed by a comparatively modern bridge of five arches, but there is record of a bridge at Barrow as early as the end of the fourteenth century. According to the *Quorndon Records*, by George F. Farnham, an inquest was held

29. ANSTEY BRIDGE.

30. FLEMING'S BRIDGE, BOTTESFORD.

31. WAKERLEY BRIDGE.

32. DUDDINGTON BRIDGE.

in August 1398 at which the verdict was given that
'John Wadyngham, late Vicar of the Church of
Barrow-on-Sore, wilfully and feloniously drowned
himself . . . in a certain water called Sore at the
bridge of Barrowe.'

The bridge leading from Cotes to Loughborough *Cotes*
is modern except for one arch, which may be part of *Bridge*
the bridge shown by Nichols. According to him
the bridge had been recently 'altered by putting
twelve arches into six, so that it now contains only
seven arches.' An inquisition was taken at Nottingham in 1333 with regard to the repair of the bridge
between 'Loughteburgh and Cotes' to decide who
was responsible for its repair.

There is now a girder bridge at Stanford-upon- *Zouch*
Soar, but, about three miles below, the bridge shown *Bridge*
by Nichols as 'Zouch Bridge' still stands as built in
the year 1793. The three arches are segmental in
shape, and every alternate voussoir projects beyond
the arch-ring. The width between the parapets is
slightly under 13 feet, and the total span measures
38 yards. The vaults of the arches are built of brick.

The last bridge across the river Soar before it joins *Kegworth*
the Trent is at Kegworth. This is said to have been *Bridge*
built in the year 1785. The five segmental arches
have projecting voussoirs, and the roadway has been
increased by carrying the footpaths on iron brackets.
At an inquisition taken in 1316 with regard to the
building of a bridge at Kegworth it was decided that
'no one is bound to construct it.' The evidence stated
that 'Henry de Sutton, parson of Leke, bequeathed
£10 and 12 oaks, of the value of 4s. 6d. each, to the
bridge of Kegworth in the confines of Leicester and

Nottingham . . . but nothing was done.' Pontage was granted in February and July 1316 and again in May 1318, in the latter case to Gervase de Clifton and two others. In August of that year a commission to audit their accounts was appointed, which was followed by a further commission in February 1321.

THE RIVER DEVON

Fleming's Bridge, Bottesford

This river, with its tributary the river Smite, drains the district south of Newark lying to the east of the Fosse Way. A few miles below Belvoir Castle it passes through the village of Bottesford. Here, a short way from the brick bridge on the main road, the Devon is crossed by an unusually attractive footbridge (Fig. 30) having two ribbed arches. This style of building was much favoured in the fourteenth and fifteenth centuries, but documentary evidence places the date of this bridge between the years 1581 and 1620. It was built by Samuel Fleming, Rector of Bottesford, who, according to Nichols, 'being once in danger to be drowned in crossing the river, he vowed that for the future no man should there run the same risque.' He made good his vow by erecting this stone bridge, which was named after him. The two arches are segmental in shape and each has three chamfered ribs. They span a total distance of about 20 feet, and the present width between the parapets is 5 feet. This is about 2 feet more than when the bridge was first built, and the widening was carried out on the downstream side by building out on to the top of the cut-water.

The river Smite, which joins the Devon about five miles north of Bottesford, is crossed by many bridges, but none of them is of any special interest. Smite Bridge in Colston Bassett, which has three segmental arches with projecting keystone, may date from the end of the eighteenth century. Most of the others are comparatively modern brick bridges, as are those over the river Devon.

The brick bridge carrying the Fosse Way across the river Devon at the southern end of Newark, now called Devon Bridge, was formerly known as Markham Bridge. The one near the Castle is shown on the 6-inch Ordnance map as Trent Bridge, although, as mentioned in Chapter I, the stream under this bridge is now considered part of the river Devon. Cornelius Brown in his *History of Newark*, published in 1904, stated that a branch of the Trent formerly passed close to the Castle, and only a small stream then ran through Averham, Kelham, and Muskham. The family of Suttons made a cut from the Trent near Farndon which took away practically the whole of the water. This affected many mills, and the Court ordered the owner of Averham to construct and maintain a weir in order to regulate the stream.

The Pipe Rolls for the year 1169 record the expenditure of 5s. 3d. ' for work on the bridge of Neworc,' and of 3s. 9d. the following year, and pontage for three years was granted in August 1346. The bridge was swept away by flood during the fifteenth century, and in the year 1486 the Bishop of Lincoln, ' bi way of almes and charite,' granted 100 marks for its repair. An agreement was also made by the ' Alderman and his brethren ' with Edward Downes, carpenter of Worksop, to build a bridge of ' good and

Bridges in Newark

sufficient oke to have 12 arches (spans?) and over the said arches rails on both sides of the bridge.' Many fifteenth-century wills included legacies for the fabric of this bridge, and in the year 1597 Queen Elizabeth granted 'three score timber trees' for Trent Bridge from the Forest of Sherwood, and also 'fortie for Markehall (Markham) Bridge.' At an enquiry held at Newark in 1627 a verdurer of the Forest said he remembered the delivery of these trees, and John Owsle, a carpenter, said that the date 1597 'was engraven upon a poast of the said bridge.' The repairs were still in hand in 1633, but fortunately there was a good ford near by, which was 'passable the greater part of the winter for horses and coaches.' The bridge was completed, however, in 1636 in time for the visit of Charles I.

The present bridge of brick and stone was built in 1775 and widened, by corbelling out the footways, in 1848. It has seven segmental arches and pilasters over the cut-waters. An engraving reproduced by Cornelius Brown shows the earlier bridge as having nine piers, seven of stone and two of timber, the latter being placed in the middle of the bridge.

CHAPTER IV

THE RIVERS AND BRIDGES OF LINCOLNSHIRE AND RUTLANDSHIRE

THE northern part of Lincolnshire is drained by many small streams, some flowing north into the river Humber, others direct into the North Sea. Few of the present bridges are of any interest, but several of their predecessors were mentioned in mediæval records.

The Patent Rolls for the year 1359 record a grant of pontage for a period of three years to the bailiffs and goodmen of 'Southferyby' for the repair of their bridge, and Nordyke Causeway, 'part of the King's Highway from Boston to the Humber,' was mentioned on several occasions. About the year 1263 the Abbey of Ravensby was ordered to make the necessary repairs. It is doubtful, however, if anything was done, as an entry in the Patent Rolls for February 8th, 1316, records the fact that a jury found 'the abbot and monks of Revesby are not bound to repair the bridge and causeway of Nordike.'

Northdyke Bridge, as it is now called, was rebuilt at the end of the nineteenth century. Only two arches of the earlier bridge were visible, but three more arches were found buried in the causeway leading to Stickney. According to the *Lincolnshire News and Queries* (vol. iii), part of a crucifix was also found, which was sent to Lincoln.

A grant was made in October 1314 allowing the

Northdyke Bridge

construction of a causeway from Louth to Cockerington, but, according to Howlett, the bridge of Louth was rebuilt towards the end of the eighteenth century. In his *Selection of Views in the County of Lincoln* he gives an illustration showing a bridge with three semicircular arches having chamfered voussoirs.

THE RIVER ANCHOLME

This river, the largest in North Lincolnshire, rises about seven miles north of Lincoln. From Bishopbridge, where it is joined by the river Rasen, it has been canalised for the greater part of its course.

West Rasen Bridge

At West Rasen this tributary is crossed by a delightful mediæval packhorse bridge, the only one of the type in these parts, although there is a small single arch bridge at Utterby which may be considered ancient. West Rasen Bridge, also known as the Bishop's Bridge, as it is supposed to have been built by Bishop Dalderby early in the fourteenth century, has three segmental arches spanning a distance of nearly 20 yards. The width is only 4 feet.

Glanford Bridge

The crossing of the river Ancholme about ten miles north of Bishopbridge was formerly known as Glanford Bridge, but the place is now called by the simple name of 'Brigg.' The 'Bridge of Glaunford' was mentioned as early as the year 1256, and it is recorded in the Roll of the Wapentake of Yarborough that during the reign of Edward I 'Gilbert Nevil had at Glandford Bridge a market on Thursday and also the toll of all merchandise sold or bought and of carts passing over the said bridge.'

In December 1312 a commission was appointed to

LINCOLNSHIRE AND RUTLANDSHIRE 57

deal with the complaint that 'the silting up of the river Ancolne, Co. Lincs, between Bishop Bridge and the bridge of Feryby towards the Humbre had caused the people along its banks to the bridge of Glannford to incur serious loss and danger.' A further enquiry was appointed in July of the following year.

In 1375 Glanford Bridge was said to be broken by default of the lord of Kettleby, and it was then stated that it formerly was only a foot-bridge of planks, but that about the year 1300 the Abbot of Thornton 'of his alms and charity, because he was born there, first repaired it with stones, and since then, when the bridge had decayed by the passage of time, it had been repaired by the alms and charity of men of the country.'

THE RIVER WITHAM

The river Witham rises on the borders of Leicester, Rutland, and the extreme south-western end of Lincolnshire, and after reaching South Witham flows due north until it reaches the city of Lincoln. The present bridges above Lincoln are uninteresting, and it is doubtful if any of them are more than a hundred years old. A bridge over the Witham at Marston was mentioned in a Presentment of 1367 when the responsibility for repair of the adjoining causeway was in dispute.

Until the year 1906 a mediæval bridge existed at Claypole having two pointed arches each with four ribs. Fortunately two excellent illustrations of this bridge are published in the *Proceedings of the Society*

Claypole Bridge

of *Antiquaries* (vol. xx), and from these it appears that both the ribs and arch-rings had unusually wide chamfers.

The Patent Rolls record that in February 1316 a commission was appointed to view Oldehebrigge, between Newark and Claypole, as it was reported that it had become dilapidated and that the stones and timber had been carried away. Fifteen years later a petition was presented by the men of Claypole stating that in the past when they wished to go to Newark 'they were wont to pass with horses, carts, etc., over the bridge of Oldeebrigge,' but this being broken down, 'they were obliged to use a new bridge built on land belonging to the Bishop of Lincoln.'

The commission appointed to consider the matter ordered that a bridge should be built 'where the ancient bridge crossed the Witham below Claypole at the west end of Claypolsouthengedykes,' and that the newly built bridge on the land of the Bishop of Lincoln should be removed.

River Till   In Lincoln the Witham is joined by the river Till, which drains the district north-west of Lincoln. Most of its bridges are built of brick and are of no great age, but Till Bridge, which carries the Roman road from Marton, is on or near the site of an ancient one which in the year 1357 was said to have been 'utterly overthrown.'

Since the time of Henry I the Till has been joined to the river Trent near Torksey by a channel called the Foss Dyke. In 1375 it was stated that this had been stopped up for a period of thirty years and the south part of 'Fossebrigge by Torksey is destroyed.'

It was a stone bridge in the sixteenth century according to Leland, but a few years ago the whole bridge was reconstructed.

Leland stated that 'the Ryver of Lincoln breking into 2 Armes a very litle above the Town passeth thorough the lower Part of Lincoln Town yn 2 severalle Partes of the South ende of the Towne very commodiusly, and over eche of them is an archid Bridge of Stone to passe thorough the principal Streate. The lesser Arme lyith more Southly, and the Bridg over it is of one Arch ... Gote Bridge to passe over the lesser Arme, Highe Bridge over the great arme.' He also mentioned that 'High Bridge hath but one great Arch, and over a pece of it is a Chapelle of S. George.'

Bridges in Lincoln

High Bridge is said to have five arches, but only one can now be seen, as in Leland's time. It has a span of about 22 feet, and the middle part, which has barrel vaulting and five ribs, is supposed to date from the year 1160. In 1235 it was widened on the eastern side to carry a chapel, which unfortunately no longer exists, and in 1540 the opposite face was extended and houses were built upon it.

In his will of 1391 John de Sutton Senr. of Lincoln bequeathed to his servant John de Brisley 'my two shops which I have on the great bridge of Lincoln.' A 'beautifully ornamented' obelisk was erected on the east side of the bridge in 1765.

Captain Armstrong in his map of Lincolnshire, published in 1778, shows no bridge between Lincoln and Tattershall, and even at the latter place the only means of crossing in Leland's days was by ferry. The present bridge at Tattershall is quite modern, as are most of those across the river Bain, which joins the Witham about a mile below Tattershall Bridge, close to the Dogdyke Ferry mentioned by Leland. A mile farther south is the junction with another river, the Slea, which on its lower reaches is called the Kyme Eau. There is nothing particular to note about any of the present bridges over this stream, but Ogilby recorded that there was a stone bridge at Sleaford, and in 1393 a jury found that the bridge called Mareham Claybrigge, between Sleaford and Threckingham (six miles south of Sleaford), was broken by default of the Prior of Sempringham.

Boston Bridge

The last bridge across the Witham is at Boston, where a bridge existed as early as the thirteenth century. In March 1305 pontage was granted to John, Duke of Brittany, for a period of three years, and in May of the same year to William de Ros of Hamelak, for five years, on the understanding that it did not interfere with the grant to the Duke of Brittany. Three years later a grant of pavage and pontage was issued to the town of Boston, and in 1313 the other two grants were extended. It seems a strange plan to allow three concurrent grants, and there must have been considerable confusion in dividing the spoil.

The town received a further grant in June 1319, but in July 1331 the name of William de Ros again appears, the grant being for 'his moiety of repairs.' The interesting schedule of the customs demanded is given by Mr. P. Thompson in his *History of Boston*.

In May 1358 pontage was granted to the bailiffs and good men of Boston for the repair of the bridge, 'now in a dangerous state,' and the following month to 'the King's son John, Earl of Richmond,' in both cases for a period of five years.

As supplies of suitable material were not available in Lincolnshire, a writ was issued in August 1358 charging the Earl of Richmond 'to hire crayers and other ships from Sandwich, Dover and Wynchelse to bring timber from Sussex,' and like instructions were issued to the 'bailiff of Hastyngges.'

According to Allen, a new bridge was built early in the sixteenth century, but this fell about the year 1556. Its successor lasted till 1629, when it was replaced by one having a stone gateway. This bridge, evidently the one seen by Ogilby, was taken down in 1742. In 1802 the erection of an iron bridge was begun, having a single arch of 86 feet span with a breadth 'over the cornice on each side' of 39 feet. It was designed by Rennie and cost nearly £22,000.

THE RIVER WELLAND

For a short part of its course the river Welland forms the boundary between the counties of Leicester and Northampton, but after reaching Rockingham it forms the south-eastern border of Rutland. The great Bridge of Market Harborough was mentioned in the will of William Neel, who died in 1439. The bridge that existed in the seventeenth century had, according to Ogilby, six stone arches. By the end of the following century it appears to have gone, as the only one shown on a map of 'Market Harborough in 1776' is shown as a 'Chain Bridge.' This chain bridge was referred to in an Act of 1721, and must have been one of the earliest of the type.

The bridge of Great Bowden benefited by the legacy of 6s. 8d. from William Southerey, whose will was proved in 1523, and the bridge at Little Bowden was described by Faden in 1791 as a 'bridge of four arches for horses and carriages leading into Harborough.' None of the existing bridges, however, are of interest, nor is Welland Bridge, four miles farther north. Nichols records that a bridge was built there in 1678 having two arches, which in 1810 was replaced by a brick bridge with one large and three small arches. The use of this bridge was only permitted when the ford was impassable owing to frost or flood.

In Melbourne there is still a packhorse bridge across a tributary of the Welland, and Leland, when travelling south from that village, crossed over the Welland, which he described as 'no great streame.'

The bridge on the track from Middleton to Drayton contains part of an earlier stone structure, but the present vault is of brick. The bridge leading north from Rockingham bears the date 1755, but its three segmental arches have been widened on both sides. This crossing was of importance even in early times, and indulgences were granted in the years 1219, 1226, 1229, and 1230, to those contributing towards its repair. The bridge at Caldecott over the Eye Brook, a mile or so farther north, is built of stone, with projecting keystones, but with brick parapets. It is, maybe, the one described by Laird, early in the nineteenth century, as a 'small rural bridge of one arch.'

Middle Bridge near Gretton is of the girder type, while the one at Harringworth is built of brick with

33. KETTON BRIDGE.

34. CHURCH BRIDGE, EMPINGHAM.

35. Deeping Gate Bridge.

36. Trinity Bridge, Crowland.

a large recess on the upstream side. The bridge was shown by Ogilby in the seventeenth century as having two stone arches.

Close to Wakerley Station the Welland is crossed by a fine mediæval bridge (Fig. 31) with five pointed arches. These have double arch-rings, both of which are chamfered. The parapets are of later date, and the bridge has been widened by about 2 feet, with segmental arches, on the upstream side. The total span is 27 yards and the width between the parapets is now 12 feet, and from an inscription 'T.S. 1793,' it appears that the widening was carried out at the end of the eighteenth century. On each side of the bridge, over one of the arches, is a carved head projecting beyond the face of the masonry. Unfortunately there appears to be no historical information about this bridge, which probably dates from the fourteenth century. *Wakerley Bridge*

Another ancient bridge (Fig. 32) crosses the Welland at Duddington, about two and a half miles north-east of Wakerley; but in this case the downstream face is completely disfigured by the widening carried out in blue brick and stone in 1919. The upstream parapet is also modern, but on this side the four pointed arches, with chamfered arch-rings, remain. The span is 22 yards, and there is now a roadway 16 feet wide in addition to a footpath 3 feet in width. In the year 1380 it was reported that Duddington Bridge 'was broken.' *Duddington Bridge*

The next bridge over the Welland, named Collyweston Bridge, is also of interest, having three pointed and three segmental arches. The latter were evidently built early in the seventeenth century, as *Collyweston Bridge* the date 1620 is carved on the upstream side of one of them. The pointed arches may also have been rebuilt; they certainly are not as ancient as those of Wakerley Bridge. The total span is 36 yards with a roadway 15 feet in width.

Bridges in Stamford

The bridge in Stamford shown by Ogilby as 'Welland Bridge' was apparently on or near the site of the present one. He also showed a road, presumably with a ford, crossing the river a short distance upstream where there is now a comparatively modern bridge which Burton, in 1846, called Water-furlong Bridge. At that date the Welland Bridge was 50 yards long and only 11 feet wide at the north end, but 14 feet in width farther south. The present bridge, having a width between the parapets of about 30 feet, was built in 1849 at the cost of £8,500 according to Walcott. Whellan, however, gives the figure £12,000. It has three segmental arches on very massive piers, while the earlier bridge had four arches.

A bridge existed at Stamford in 1149, and a Charter from Richard I to the Abbey of Burgh mentioned land in Stamford 'beyond the bridge.' A hospital and chapel dedicated to St. John the Baptist and St. Thomas of Canterbury was built in 1174 at the bridge-head on the south side. The north end of the bridge was broken by flood in the year 1570. Burton in his *Chronology of Stamford* also shows three bridges in the meadows: 'Broadeng 16 chains east of the Wash,' Lammas over the north arm of the Welland at the bottom of Castle-dyke, both of timber, and 'George Bridge' of timber on stone piers. There was also one across the King's Mill Stream leading from Austen Street to Nortons

LINCOLNSHIRE AND RUTLANDSHIRE 65

Terrace, bearing the name of Melancholy Walk Bridge. Walcott gives a picture of the bridge by Burghley House, showing three semicircular arches and balustrades, each of which terminated with a stone lion.

In Stamford, close to Welland Bridge, this river is joined by the Chater, and a mile or so farther east by the Gwash. The only bridge of note over the Chater is the one at Ketton (Fig. 33), which has three pointed arches, widened on the downstream side in 1849. Leland noted that the bridge had six arches, and it is possible that the other three still remain in the causeway leading to the bridge, but hidden by the massive abutments built evidently when the bridge was widened. The roadway is now 13 feet wide and the total span over the three arches measures 16 yards. *River Chater*

The river Gwash rises about four miles west of Oakham. Ogilby showed a stone bridge at Manton, but now the first bridge of importance is the one at the southern end of Empingham, called Church Bridge. A bridge existed here in 1684 according to the map in James Wright's *History of Rutland*, possibly the one (Fig. 34) still in use. The three arches are nearly semicircular in shape, with triple arch-rings, and the cut-waters are exceedingly massive. The roadway is 13 feet in width, and there are recesses for foot-passengers over the cut-waters. The total span is 22 yards. *River Gwash*

Ermine Street crosses the river Gwash at Great Casterton, formerly known as Bridge Casterton. The present bridge is of no interest, but in the sixteenth century the river was crossed by a stone bridge with three arches which Leland called 'Castleford Bridge,' although a century later Ogilby marked it as 'Casterton bridg.'

Two bridges cross the river near Ryhall. The one carrying the road from Stamford to Little Bytham was probably built late in the eighteenth century, but the bridge near the church is at least a hundred years older, the three semicircular arches, built in two orders, being typical of the seventeenth century. They span a distance of 15 yards. On the upstream side a footway, about 2 feet in width, has been built across the tops of the cut-waters. Iron railings take the place of the usual parapets, and the roadway is 13 feet in width.

The last bridge over the Gwash carries the road from Stamford to Uffington. It is built of stone as in the days of Leland, but has been widened to such an extent that it is difficult to tell its age.

Uffington Bridge

Between Uffington and Barnack the Welland is crossed by another bridge with semicircular arches, the arch-rings being built in two orders. The total span of the three arches is 30 yards. There are two cut-waters on each side which probably once had recesses at the road level, but the parapets, which are 13 feet apart, have been rebuilt. The bridge of 'Offynton near Stanford' is mentioned in several of the thirteenth and fifteenth century Uffington Charters dealing with the Convent of Newstead.

Lolham Bridges

The bridges at Tallington and West Deeping were probably built early in the nineteenth century, but a short way south of the latter a long causeway, called Lolham Bridges, contains several seventeenth-century arches. On one of the cut-waters between the arches immediately to the north of the level-crossing is an inscription giving the date 1641, while on another arch still farther north, which has projecting keystones, is carved the date 1790. The arch-rings of these arches are quite plain, but those of the four semicircular arches south of the crossing are chamfered. Evidently the whole causeway was repaired during the second quarter of the seventeenth century, as in the year 1621 Sir George Manners wrote that 'the decayed bridge at West Deeping, th' ordinarie roade way from Lincolne to London, are verie fowle and dangerous.'

The present bridge at Market Deeping is scarcely more than a century old, but just over a mile below, at Deeping Gate, a fine bridge (Fig. 35), with three segmental arches, crosses the Welland. These are nearly semicircular in shape and have triple arch-rings built in three orders, an unusual feature, but found in several bridges built in the seventeenth century. The middle arch has a brick vault and appears to have been partially rebuilt. The roadway is 13 feet wide, but there are recesses over each of the cut-waters. *Deeping Gate Bridge*

Below Deeping Gate and Deeping St. James the Welland enters the Fen district, often dividing into several streams, and after reaching Crowland it strikes due north.

One branch of the Welland formerly flowed through the village of Crowland, where it was joined by another stream. The spot where they met appears to have coincided with the junction of three paths, and here the monks of Croyland, as it was formerly spelt, erected the famous Trinity Bridge. *Crowland Bridge*

The streams have disappeared and their beds have been filled up, but the bridge itself (Fig. 36) still remains. It might be described as consisting of three 'half-arches' built 120 degrees apart, each carrying a footway with steps. Each part is about 6 feet in width and has three ribs, and the points where the middle ribs enter the present ground level form an equilateral triangle with sides 20 feet in length. In the middle of the nineteenth century Canon Moore, F.S.A., made a careful examination of this bridge and wrote a paper on the subject, in which he stated that 'the hollow chamfer of the groining ribs, the wave moulds on the face of the arches, the three-quarter hollow in the re-entering angle between the two members, and the shape of the arches indicate erection between 1360 and 1390.' A note in Bowen's edition of Ogilby's *Road Book* states that Crowland was built 'on Piles like Venice (if wee may make ye comparison) consisting of 3 Streets which have a communication by a Triangular bridge: it is so remote from Pasture that ye Inhabitants are obliged to goe a milking by water in little boats called Skerrys wch carry 2 or 3 persons at a time.'

River Glen   About four miles below Spalding, where only timber bridges existed before the nineteenth century, the Welland is joined by the river Glen. Most of the bridges over this stream are comparatively modern, Manthorpe Bridge bearing the date 1813 and Edenham Bridge that of 1831. The bridge at Essendine may date from the seventeenth century, its semicircular stone arches being typical of that period. Wooden railings, giving a slightly wider roadway, take the place of the usual parapets. The total span of the three arches is 10 yards and the overall width of the bridge is about 15 feet.

The two branches of the river Glen meet near Wilsthorpe, and a short distance below is situated Kate's Bridge, which carries the road from Market Deeping to Bourne. A presentment of 1349, concerning a marsh dyke called 'Edyk,' makes mention of 'Katebrigg,' and 'the common fishery in the waters of Catebrigg' was the subject of an agreement made by Richard, Abbot of Croyland, in the year 1245. In 1674 a bridge was shown by Ogilby as 'Ket Bridge of Stone over Boston Drain,' but from its position on the road north from Market Deeping it is certainly the one now known as Kate's Bridge.

Ogilby showed the bridge at Surfleet as of 'wood,' here calling the river the Boston Dike, while the name Bourn Eau is used in the complaint, lodged in the year 1375, that 'the bridge by Thomas Dode's house in Surfleet,' being broken, was choking the river and causing floods. 'Toftbrygge,' mentioned in another presentment of the same date as a footbridge, between Gosberton and Surfleet, was perhaps the one over the stream now called the Risgate Eau.

The last bridge over the Welland, which carries the road from Holbeach to Boston, is at Fosdyke. In the year 1794 an Act was passed authorising the building of a bridge, but the promoters were unable to raise the money and the work was delayed for nearly eighteen years. The bridge then built was designed by John Rennie. According to W. Marrat, it had 'eight openings and nine piers, each of six oak trees.' The total length was 100 yards and

Fosdyke Bridge

the width measured 22 feet. The middle opening had two draw-leaves. This bridge remained in use until 1871.

Holland Causeway, leading from Spalding to Donington, via Gosberton, extended to Bridge End where the convent of St. Saviour formerly stood. It contained thirty separate arches, each 10 feet in width and 8 feet high, and was a continual source of trouble.

In February 1307 pontage was granted to the Prior of Sempingham for the maintenance of the bridge and causey of Holland, and this was renewed in November 1320. In 1321 it was ordered that the Prior of St. Saviour should undertake repairs, but ten years later thirteen arches were stated to be out of repair. Further grants were made in September 1334 and in January 1347, in the latter for a space of six years, and in both cases the Prior of St. Saviour was made responsible. A commission was appointed in August 1352, as the bridge was said to be broken down, and the following year the Prior obtained a further grant, which was extended to 1357 for an additional five years. It was again prolonged in 1363 and in May 1368. During this period there had evidently been some trouble, as a commission was appointed in November 1366 to audit the Prior's accounts, and pontage was granted about the same date to Alan Walter for three years. This grant was revoked in favour of the Prior in February 1367, and the Prior obtained further grants in June 1373 and July 1375. At the latter date it was stated that the causeway was in ruins. Twelve further grants maintained the rights of pontage until the year 1420.

37. EVERDON BRIDGE.

38. DESBOROUGH BRIDGE.

39. GEDDINGTON BRIDGE.

40. DITCHFORD BRIDGE.

CHAPTER V

THE RIVER NENE

STUDBOROUGH HILL, three miles south-west of Daventry, forms the watershed of three rivers. The Leam flows towards the north into the Avon and eventually reaches the river Severn, the Cherwell runs south into the Thames, whilst the Nene, which rises on its eastern slopes, discharges into the Wash.

Throughout its course the Nene is fed by numerous tributaries, few of which appear to have any names, although, strangely enough, most of their bridges bear names. Few of them, however, are of any archæological interest, the majority of the existing structures dating from after the end of the eighteenth century.

About three miles south-east of Daventry and a short distance from Great Everdon Church the road to Farthingstone crosses a tributary, called by Baker 'the Fawsley Water,' over a mediæval bridge (Fig. 37) having two pointed arches. Each of these has three chamfered ribs. The total span of the two arches is 8 yards and the width between parapets measures $11\frac{1}{2}$ feet. On the upstream side there is a recess over the cut-water, while on the other face the bridge is strengthened by means of a flat buttress. Bridges, in his *History of Northampton*, records that 'the manor of Little Everdon was granted to Cardinal Wolsey,' and also stated that 'an half-yard land and a close were formerly given to this parish by a lady, whose name is now forgotten, for the re-

Everdon Bridge

paration of the Church, the bells and the mill end of the bridge,' and that 'these lands are in the hands of trustees and yearly let out to the best advantage. The annual profit arising from the close is somewhat better than five pounds.'

Another bridge with pointed arches can be seen three miles north-east of Daventry, close to Whitton Mill. The arches are of much later date than those of Everdon Bridge, and the entire structure above the arch-rings, and also the cut-waters, have been rebuilt with blue bricks, forming a ghastly travesty of an ancient bridge.

Bridges in Northampton

In Northampton the Nene is joined by a tributary coming from Chapel Brampton. The road to Floore crosses this stream at West Bridge, which was described by Leland as 'fairly arched with stone.' He mentioned St. Thomas Bridge at Northampton, but gave no details concerning it, and was probably referring to the one by which the road to Stony Stratford crossed the main stream of the Nene at the southern side of the town. During the Civil War both were made into draw-bridges, and in 1663, according to Bridges, a 'prodigious flood . . . burst the west bridge and forced away the chief arches of the south bridge.' When the latter was rebuilt 'the two arches were turned into one large one.' Ogilby, towards the end of the seventeenth century, showed a stone bridge with '6 Arches,' but the present one is of much later date.

St. Peter's Bridge, two miles east of Northampton, is an iron structure, built in 1842, while Billing Bridge, three miles below, is built of brick and stone. The latter crossing was evidently of importance in the Middle Ages, as in the fourth year of the reign of Edward I the jurors presented 'that Sir Roger de Wauton had withheld from the King and the bailiff of Northampton for four years past the customs and tolls due at Billing bridge.' Another jury complained that he took a toll of '2s. for every pair of millstones going to Northampton.'

No bridges are shown on T. Jeffrey's map of the County of Northampton (1779) between Billing Bridge and the one, a mile south of Wellingborough, called Long Bridge. This is now a girder bridge, but in the seventeenth century there was one of timber. Formerly it was known as Brampton Bridge.

A short distance below Long Bridge the river Ise joins the Nene. It rises about five miles south-west of Market Harborough and, flowing east as far as Geddington, turns due south, skirting the eastern side of Wellingborough. *River Ise*

The small bridge about a mile south of Great Oxendon is said to be on the site of one which formed one of the boundary points of Rockingham Forest in the time of Edward I, while the 'bridge of Aringworth,' now called Arthingworth, served the same purpose in a perambulation made of Whittlebury Forest in the twelfth century. The present bridge at Arthingworth has two almost semi-circular arches and may date from late in the seventeenth century. In its original state the overall width was scarcely more than 5 feet, but the bridge has been widened in brick to give a roadway of 18 feet.

Newbottle Bridge, two miles below, possibly also of the same age, has suffered a like fate. It has, however, only one arch.

74 ANCIENT BRIDGES

Desborough Bridge

The bridge on the road from Desborough to Rothwell bears the date 1780. The two arches (Fig. 38) are semicircular in shape, as are the cut-waters. The width between parapets is 18 feet, and the upper part of the bridge appears to have been rebuilt.

On Ruston Bridge is carved the date 1829, but Barford Bridge, three miles north of Kettering, was rebuilt in 1918. A stone bridge was shown here by Ogilby, but he did not state the number of arches. In the year 1329 the bridge of 'Bereford,' leading from Kettering towards Staunford, was reported as 'thrown down and broken.'

Geddington Bridge

In Geddington the river Ise is crossed by an unusually massive bridge. Three of its four arches (Fig. 39) are pointed in shape, but at least one of these appears to have been rebuilt, and the southern arch has been repaired with blue bricks. The keystone of the middle arch, which is semicircular, bears the date 1784. The cut-waters are immense, with correspondingly large recesses for foot-passengers, but these were necessary as the parapets are less than 11 feet apart. The total span is 29 yards. A by-pass bridge has recently been built a short distance upstream of this ancient bridge.

Ditchford Bridge

Two miles east of Wellingborough, and close to Ditchford Station, the river Nene is crossed by Ditchford Bridge (Fig. 40), which has six arches across the main stream and three in each of the approaches. The arch-rings are double and built in two orders, while the six arches span a distance of 50 yards. The width between the parapets is slightly less than 14 feet.

As at Geddington, there are very massive cut-

41. IRTHLINGBOROUGH BRIDGE.

42. BRIGSTOCK BRIDGE.

43. FOTHERINGHAY BRIDGE.

44. WANSFORD BRIDGE.

THE RIVER NENE

waters, and on the upstream parapet of the second recess, from the northern end, is carved on its northern face the crossed keys of the Abbey of Peterborough, and on the other a St. Catherine's wheel having eight spokes. The bridge was restored in the year 1924, which accounts for its new appearance. A bridge was in existence here early in the fourteenth century, as in the year 1330 the 'bridge and causey of Dicheford' were presented as being ruinous to passengers.

Between Irthlingborough and Higham Ferrers the river Nene is crossed by a magnificent mediæval bridge (Fig. 41) with nineteen arches spanning a distance of 90 yards. The majority of the arches are pointed in shape, dating probably from the fourteenth century. This bridge was widened in 1922 by building out on to the top of the upstream cut-waters, and has now a roadway 17 feet in width. The ancient arches are about 10 feet in overall width, and several have narrow chamfered ribs. Those at each end of the bridge are plain, but it is possible that originally they also were ribbed. On the fourth downstream cut-water from the eastern end is carved the date 1668, and on one of the corresponding cut-waters on the upstream side the crossed keys again appear. The bridge and highway at Irthlingborough were the subject of a bequest in the will of John Pyel of London, who died in the year 1382, but otherwise nothing appears to be known concerning the history of this interesting bridge.

Irthlingborough Bridge

Concrete has taken the place of timber at Denford, a mile and a half south of Thrapston, but the bridge leading from Islip to Thrapston is of stone with nine

Thrapston Bridge

segmental arches. The roadway is 24 feet in width. Indulgences were granted in 1224 and 1226 to those contributing towards the maintenance of Thrapston Bridge, and pontage was granted for its repair in June 1369, June 1373, May 1382, June 1388, February 1393, and February 1411.

Leland described the bridge as having eight stone arches, as recorded by Bridges towards the end of the eighteenth century, when the tolls charged were twopence per waggon and a halfpenny a horse. Thrapston Bridge was one of the many that were damaged by the flood of 1795.

Harper's Brook

A mile and a half below Thrapston a small stream called the Harper's Brook joins the river Nene. Most of the bridges over this stream are modern, but one at the western end of the village of Brigstock (Fig. 42) is of considerable age. Each of its semicircular arches has double arch-rings, which are chamfered, while the upstream cut-water is extremely massive. The total span of its two arches is 12 yards, and the roadway is 14 feet in width.

Between Aldwinkle All Saints and Thorpe Waterville the Nene is crossed by a stone bridge, known as Brancey Bridge, which has three plain semicircular arches, bearing the date 1760. The bridge on the western side of Lilford Park is, however, of later date. The route given by Ogilby from St. Neots to Oakham crossed the Nene at this point by a stone bridge, and he showed the adjoining village as 'Pilkinton als (alias) Pilton.' There was also a stone bridge here in Leland's days. Neither writer recorded the number of arches, but Bridges, early in the eighteenth century, gave it as ten; the seven nearest to Lilford Hall being 'all walled in' (*i.e.*, with parapets), the other three, which belonged to the lordship of Pilton, being 'only railed.' A new bridge, having fluted pilasters, was built in 1796.

Leland remarked about the bridges at Oundle that there was one with 'five great Arches and two smaul' on the road to the south, and another which he crossed 'going oute of the Toune end of Oundale towards Fodringeye.' This one he called the North Bridge, and 'gessid that there were about 30 Arches of smaul and great that bare up this Cawsey.'

Bridges in Oundle

Bridges described the former as 'Barnwell Bridge,' and also mentioned 'a stone bridge for horses over a fordable brook' on the way to Stoke Doyley.

According to the *Victoria County History for Northampton*, Abbot Godfrey in 1329 claimed a 'through toll' at Oundle, alleging that 'in former days there was no common way through Oundle on account of the inundation of the waters,' and that this toll was granted 'for licence to pass through the abbot's land and to make two bridges (at the cost of the county) on this soil.'

Pontage was granted in February, 1360, to the bailiffs and good men of Undle for the repair of 'Asshetonbrigg near Undle,' presumably the North Bridge, a privilege which was extended almost continuously until early in the fifteenth century.

The North Bridge was rebuilt in 1912, and during a former rebuilding, in 1835, a tablet from a still earlier bridge was found which bore the inscription 'In the yere of oure Lord 1570 thes arches wer borne doune by the waters extremytie. In the yere of our

Lord 1571 they were bulded agayn with lyme and stonne. Thanks be to God.'

The southern bridge, now sometimes called Crowthorp Bridge, has six semicircular arches, with projecting keystones, dating perhaps from the end of the eighteenth century.

Fotheringhay Bridge

The present bridge over the Nene at Fotheringhay (Fig. 43) has four segmental arches spanning a width of 30 yards. These are built in two orders, the inner one being chamfered. The width between the parapets measures 12 feet, and the unusually large keystones project beyond the face of the bridge. It was probably constructed during the second half of the eighteenth century, as when Bridges compiled his notes, about the year 1719, the bridge built by Queen Elizabeth was still standing. He described it as having 'four arches (spans ?) covered with wood and stone laid upon it, partly walled and partly railed in.' In the illustration he gave from a drawing made in 1718, it appears that it was a timber bridge on stone piers. A stone tablet, mentioned by Stukely, bore the inscription, 'This bridge was made by Queen Elizabeth in the 15 yere of her Reygne A⁰ D^{ni} 1573'; also 'God save the Queen'; while over it, in a circle, were cut the initials 'ER' with 'a knot between.' In Leland's time it was a timber bridge.

Willow Brook

The Willow Brook joins the river Nene at Elton, about two miles north-east of Fotheringhay. Its bridges are mainly small single-arch structures, some probably dating from the end of the eighteenth century. The one at Blatherwycke, which was rebuilt in 1905, bears records of several earlier bridges, the dates 1657 and 1706 being carved on a stone on one of the upstream cut-waters, while on the second cut-water on that side is the date 1826.

The bridge over the Willow Brook, a short distance north-east of Fotheringhay, has projecting keystones.

An important road evidently crossed this brook in the neighbourhood of Walcot, as in the year 1329 it was stated that 'the bridge of Walcotforth, which is a common passage for men, foot, horse, and carriages from the town of Oundle to Staunford, is thrown down and broke.' The verdict given was that 'the people of Fodryngey and Nassington ought to repair and maintain that bridge.'

Wansford Bridge, across the river Nene, has fortunately been saved from destruction or widening by the construction of a massive concrete bridge about a quarter of a mile further downstream. The old bridge (Fig. 44) has twelve arches, all nearly semicircular in shape, spanning a distance of about 100 yards. The width between the parapets is only 14 feet, and on them is carved 'P. M. 1577.' There is also the record of the rebuilding of three arches between the years 1672-1674.

Wansford Bridge

Indulgences were granted in 1221 to all travellers giving alms for the repair of 'Walmesforde Bridge,' and the Close Rolls for the year 1233 record the grant of one oak from the forest of Clive to 'the Keeper of Walmesford Bridge for the work of the bridge.'

Many grants of pontage were issued, from January 1333 onwards, for its repair, the latest being one of seven years granted in May 1444. According to Hollingshed, three arches were destroyed during

Peterborough Bridge

'the tempest' of 1571, and some of the southern arches were damaged by ice in February 1795.

An iron bridge now spans the river Nene at Peterborough. An earlier one was reported as being damaged by flood in 1795, and even as far back as the fourteenth century the bridge of Peterborough was the source of much trouble.

Bridges recorded that in the year 1307 Abbot Godfrey built a bridge here at the cost of £14 8s. od., which was, however, broken by ice the following year. Another was built which cost £18 5s. od., and early in the fourteenth century Adam de Boothby is stated to have 'repaired the bridge at the entrance of the toun,' while John Chambers, the last Abbot and the first Bishop of Peterborough, left in his will, which was proved in the year 1556, the sum of £20 'to the repair of the bridge.'

The Patent Rolls for May 1346 record an enquiry concerning the repair of the bridge of Peterborough, at which it was stated that 'none are bound to repair the bridge because there was no bridge there until Godfrey, sometime Abbot of Peterborough, as an alms and of his own will, charitably built one in the fourth year of the reign of Edward II, and kept it in repair in his time. After his death the bridge became ruinous, and remained so until the present King and the Queen, his mother, came there and, out of reverence for them, Adam, then Abbot of Peterborough, although not bound to do so, repaired the same with wood, boards and bolts for their passage only.' Pontage for a period of three years was granted in the year 1334, also in 1352, and again in January 1374.

THE RIVER NENE

Below Peterborough the river Nene now flows by artificial channels direct to Wisbech. Its old course was through Farcet, Ramsey, March, and Outwell. J. Moore, in his map of the Fens published in 1684, showed bridges at these places, also several over the Forty Foot or Vermuden's Drain and the Whittlesey Dike. The existing bridges are of no interest, but at Wistow, about three miles south of Ramsey, a small tributary is crossed by a nice stone bridge with three semicircular arches. The total span is 13 yards, and the roadway has now a width of 15 feet, the widening having been carried out in concrete. Pontage for the repair of the bridge at Wisbech was granted in October 1331 to the Bishop of Ely for a period of three years.

CHAPTER VI

THE RIVER OUSE

THE river Ouse rises near Syresham, on the borders of the counties of Buckingham and Northampton. It flows south through Turweston and the eastern side of Brackley until it reaches Evenley Park, thereafter taking an easterly course to Buckingham. Near Evenley a small brick bridge carries the road from Brackley to Oxford across a small tributary. Here in the days of Leland there was a stone bridge, with a single arch, but Bridges, in the eighteenth century, remarked upon a stone one with two arches, dividing Brackley from Evenley. The fact that bridges existed over this part of the Ouse in the thirteenth century is proved by the boundaries of the Bailewick of Buckingham, recorded in the year 1252. According to the Calendar of Inquisitions (Misc.) these extended 'from Thoty Bridge along the river called Huse up to the mill of Gravesende and so to Turveston Bridge and so to Childebridge and so to the Chapel of Luffield ... to Lillingston Dayrel and so to Thoty Bridge.' Thoty Bridge was also given as a boundary of the Bailewick of Wackefeld in the forest of Witlewood, as are also Stratford Bridge, Achelsford Bridge, and Twiford Bridge. It is difficult to identify these bridges, but Stratford Bridge was probably the one at Stony Stratford.

Bridges in Buckingham Three bridges were shown by Speed, early in the seventeenth century, as crossing the river Ouse in the town of Buckingham. All these have been rebuilt, the one on the road to Bicester, called Castle Bridge, as lately as 1924. Lord's Bridge, at the southern end of the town, has two brick arches possibly dating from the middle of the eighteenth century, and London Bridge, on the road to Padbury and Aylesbury, which has three segmental stone arches with projecting keystones, was built in 1805 by the Marquess of Bath. According to the *Victoria County History*, the adjacent iron foot-bridge replaces an ancient stone bridge known as the Sheriffs Bridge. Ogilby showed three bridges, and stated that two of them had six stone arches apiece.

About two and a half miles east of Buckingham the Ouse is joined by the Claydon Brook which, with its many tortuous tributaries, drains the vale lying south and west of Buckingham. Ogilby showed only wooden bridges on the road from Padbury to East Claydon, but the present ones are built of brick. Padbury Bridge, which carries the road from Buckingham to Winslow, is, however, of stone and, according to Lipscombe, was erected in the year 1828 in place of a bridge built in 1742. It has three segmental arches, with projecting keystones, and iron railings with stone pillars at each end.

Nothing appears to be known about the history of Thornborough Bridge (Fig. 45), which carries the road from Buckingham to Fenny Stratford across the Claydon Brook. It is by far the oldest bridge in the whole of Buckinghamshire, and possibly may be the one called Totisbrigge, in a Presentment of 1389, and described as being 'in the parishes of Thornborough, Leckhampstead, and Foxcott.' It was stated that Totisbrigge was broken, and a proposal

Thornborough Bridge

was made that 'the Hospital of St. John the Baptist at Oxford should repair the foot of the bridge and one arch on the Thornborough side, and the men of Leckhampstead and Foxcott the foot and two arches on their side.'

The present bridge has six stone arches, pointed in shape, with double arch-rings built in two orders. Two arches have four chamfered ribs. On the upstream side are three cut-waters with recesses, but downstream there are none. In place of these there are two small buttresses and a large rectangular recess carried by an extension of one of the piers. The total span is 42 yards and the width between the parapets measures about 10 feet.

Stony Stratford Bridge

Girder bridges now cross the Ouse at Thornton and at Passenham, but in Old Stratford there is a stone bridge with three arches built early in the nineteenth century. The bridge of Stony Stratford was mentioned in a perambulation made in the year 1299, and grants of pontage for its repair were made in the years 1349, 1352, 1380, and 1400. In the last case it was granted to John Blawemuster, who was described as an 'ermyt.' By his will made in the year 1520 Thomas Pygot devised his inn, called the Cock, at Stony Stratford 'to feofees for evermore to the sustentation and maintenance of the brigg of Stony Stratford.' Ogilby showed two bridges, one with five and the other with eight arches, and Bridges, early in the eighteenth century, mentioned 'a long bridge which is kept in repair by an estate settled for that purpose.' In 1834 an Act was passed to remove the bridge on account of 'its narrow, decayed and dangerous state.'

45. THORNBOROUGH BRIDGE.

46. HARROLD BRIDGE.

47. STAFFORD BRIDGE.

48. BARFORD BRIDGE.

THE RIVER OUSE

A short distance below Stony Stratford the Ouse receives the river Tove, which drains the district lying to the north of Buckingham. None of its present bridges are of interest, but Towcester, being on Watling Street, must always have been an important crossing. In 1366 Richard de Tovecestre, John his brother, and ten others, with all 'the commonalty of Toucestre,' were appointed to collect alms for the repair of the bridge near the hospital called 'Northbrughge.' Indulgences were granted by the Bishops of Salisbury, Lincoln, Coventry, and Lichfield to those contributing towards its repairs. They were also granted by the Archbishop of Canterbury.

River Tove

In Newport Pagnell the road to the north crosses the Ouse by a bridge built in three parts. The northernmost, called Lathbury Bridge, has three segmental stone arches twice widened in brick; the middle one has three brick arches on stone piers; and the part nearest the town, shown on the 6-inch Ordnance map as 'North Bridge,' is built of stone and bears the date 1787. Its single arch has projecting keystones.

Newport Pagnell Bridges

In November 1380 pontage was granted for a period of three years for the repair of 'the Northbrugge and the Southbrigge at Newportpagnell,' and the bridge was mentioned in the will of Richard Towres, who died in the year 1421.

In Newport Pagnell the Ouse is joined by the river Ouzel, sometimes called the Lovat. It rises on the north-western slopes of the Dunstable Downs, close to Whipsnade. Northall Bridge, near Billington, was reported to the Bedfordshire Sessions of 1832 as being 'narrow and out of repair,' and an

River Ouzel

estimate was obtained for building a new bridge. It was to be built of brick, the cost being £140.

The bridge between Linslade and Leighton Buzzard was reported in 1815 to be 'in decay,' and nine years later was 'washed down by a calamitous flood.'

At Fenny Stratford, Watling Street crosses the Ouzel. A royal writ was issued in 1347 giving instructions 'to cause as many bridges to be made between Leighton Buzzard and Fenny Stratford as there used to be,' and grants of pontage were issued in March 1383, July 1398, and February 1401. A cast-iron bridge known as Tickford Bridge now crosses the Ouzel in Newport Pagnell. It was described by Lipscombe in 1831 as 'of modern design and highly ornamental.'

Sherington Bridge

A mile north-east of Newport Pagnell the road to Bedford crosses the Ouse by Sherington Bridge. This massive structure has five segmental arches, with projecting keystones, spanning a distance of 66 yards. It has semicircular cut-waters over which are pilasters extending to the tops of the parapets, a type much favoured at the beginning of the nineteenth century. Part of Olney Bridge, three and a half miles farther north, appears to date from the same period. According to the *Victoria County History*, Richard Maryot, lord of the manor of Caves, in 1490 bequeathed six marks 'to the making of the arches of the brigge of Shiryngton, not now vaulted with stone, with a perpoynt (parapet?) wall upon the said arches, if they will not be made with less silver.'

Turvey Bridge

The road from Olney to Bedford crosses the Ouse between Cold Brayfield and Turvey by a long cause-

THE RIVER OUSE

way called Turvey Bridge, evidently of great age. Mention was made in the Drayton Charters (1138-1147) of 'a meadow next Turvey Bridge,' and in 1825 costs in connection with repairs to the bridge amounted to £120 8s. 1d. This included the cost of searching the records to find who had done repairs in the past, so evidently even at that date the county did not consider itself responsible for the maintenance of the bridge. Two arches on the Buckingham side were damaged by flood in 1872, and a few years ago the downstream side was widened.

Another long causeway crosses the Ouse between Harrold and Chellington, about three and a half miles below Turvey. The road is carried across the main stream by a bridge (Fig. 46) having five pointed arches, spanning about 65 yards, eastward of which are eight more, semicircular in form, which span a further 60 yards. Beyond these, and alongside the road, is a foot-bridge nearly 250 yards in length, with nineteen arches, all but four of which are pointed.

Harrold Bridge

The original part of one of the arches over the river has double arch-rings, built in two orders, both chamfered. The others, although pointed in shape, appear to have been rebuilt. The roadway is about 12 feet in width, but between the pointed and the semicircular arches it has been increased to 20 feet for a distance of about 15 yards.

The Hundred Rolls of 1279 mention a 'fishery from Harrold Bridge to Odell ford,' and Ogilby noted that the bridge was of stone. An account for repairs to this bridge was before the Sessions in the year 1765, and further work required to be done in

1806 and again in 1824, the last mentioned being to make good the damage caused by flood.

Below Harrold Bridge the Ouse strikes north-east to Sharnbrook, after which it takes a winding course towards the south. Neither Felmersham nor Radwell Bridges are shown on T. Jeffery's map of 1765, the next crossing given by him being Stafford Bridge, leading from Pavenham to Oakley and Bedford.

Stafford Bridge

The origin of the name of this bridge has been the subject of some discussion, and it is interesting to find an entry in the Common Pleas for the Bedford Eyre of 1227, published by the Bedfordshire Historical Record Society (vol. iii.), that 'Richard de Pabeham fell from a certain mare into the water of Stafford so that he died,' by which it appears that even in the thirteenth century this part of the river bore the name of Stafford. It was shown as Stafford Bridge by Ogilby in 1674.

In 1755 an estimate was presented to the Sessions for work to be done on Stafford Bridge, but in 1760 the bridge was still out of repair. A committee, appointed in 1818 to enquire who was responsible for its maintenance, decided that it was the joint responsibility of the Duke of Bedford, Mrs. Winstanley, and the county. In April 1820 Stafford Bridge sustained considerable damage by flood, and it was reported that 'ice drawn against it with great violence has forced out several stones in most of the arches.' The repairs cost over £200, including 'painting both sides of the bridge, being 417 yards long.' This presumably referred to the railings, as even now the bridge has no parapets. Further damage by flood was caused in 1825.

When last seen by the author, this bridge (Fig. 47) was in a rather dilapidated state, but two mediæval segmental arches, each having two chamfered ribs, still remained. The roadway over these arches had been widened by means of wooden beams, laid across the tops of the cut-waters, but the remaining span was crossed by iron girders. It is the only bridge of this type left in these parts, so it is hoped that efforts will be made to preserve it.

The road to Bedford from Northampton crosses the river Ouse between the villages of Bromham and Biddenham. The bridge is now called Bromham Bridge, but was described as 'the bridge of Bideham' in an entry in the Calendar of Fines for the year 1227-8, and as the 'Bridge of Biddenham' in the Pipe Roll of 1224, when the sum of 4s. was spent on its repairs.

Bromham Bridge

According to the *Annals of Dunstable*, the bridge was broken by ice in the year 1281, and one unfortunate woman was carried away on an ice-floe. She was seen passing under Bedford Bridge, four miles below, but there is no record of her having been saved. In the Clerical Subsidies of the year 1400 it was still known as the 'pontem de Bydenham,' and the Chantry Certificates (*c.* 1540) used the same name when referring to a 'Chauntry to our Lady and St. Kateryn at the foot of this bridge.' According to William Harvey, in his *History of the Hundred of Willey* (1872-1878), remains of this chapel could then be seen in the structure of the miller's house, but this unfortunately has since been destroyed.

From the Sessions Rolls it appears that Bromham Bridge was undergoing repairs in the year 1728, and in 1750 it was reported that the last arch at the eastern end required attention. A bill for masons' work was presented to the Sessions three years later, another in 1775, and a further account the next year. A new arch was 'turned' ten years later, and even in 1792 Bromham Bridge was only 'a horse and foot bridge,' according to a presentment made that year for its repair. Nothing seems to have been done, as in 1799 it was said to be 'ruinous and in decay.' An account for work done was, however, sent in six years later, so evidently the repairs were eventually put in hand.

In 1813 Robert Salmon made a plan of Bromham Bridge costing the sum of 4 guineas. He was evidently the surveyor, as in September 1814 he wrote to John Miller informing him that his contract for Bromham Bridge was completed. The following year Salmon's account for extra work on the bridge amounted to £30. No further references are made concerning Bromham Bridge until 1823, when £2 12s. 7d. was paid for bricklayers' and stonemasons' work. In the following January it was again stated to be in decay.

Bromham Bridge appears to have been considerably reconstructed, or perhaps rebuilt, early in the nineteenth century. It has now twenty-six semicircular arches and a width between parapets of 18 feet.

Bedford Bridge

Bedford Bridge and its chapel was mentioned in the Charter of Symon de Beauchamp which was granted between the years 1179 and 1194. About the year 1331 an oratory, dedicated to St. Thomas, was built upon the bridge, and the right of appointment of the chaplain, who was to collect money for the upkeep of the bridge, was claimed by the mayor. This right was violated in 1332 by the Sheriff, who appointed John of Derby and ejected the chaplain already in office. The dispute lasted for over twelve years and was eventually settled in favour of the King, as in February 1349 pontage for three years was granted to the mayor, bailiffs, and goodmen of Bedford, at the request of John de Tamworth, 'warden of the King's free Chapel of St. Thomas the Martyr on the bridge.' It was extended in February 1359 and again in March 1383.

This chapel was evidently rebuilt in the fifteenth century, as Thomas Charlton, Alderman of the City of London, bequeathed in his will, which was proved in 1452, the sum of 40s. 'towards the making of the new Chapel of St. Thomas the Martyr at the foot of Bedford Bridge.' He also left 5 marks 'towards the reparacion of the said bridge.'

In 1671, as the result of a great flood, it was reported that 'the ston house called ye Bridge house is totalee fallen down and ye rest much shaken and like to fall and ye foundacion or pile where on it stood, a great part washed away.' It was agreed and ordered that 'the prison upon the Bridge shall be rebuilt.'

The bridge, which was pulled down early in the nineteenth century, was described by James Wyatt as having 'two arches at St. Paul's end, then two gatehouses in the centre having upper chambers, then three arches at St. Mary's or the south end.' The gatehouse over the bridge was removed in 1765.

At one time a curfew had evidently been enforced,

92 ANCIENT BRIDGES

as there is an entry in the books of the Corporation which reads: 'Item yt ys ordered that the great cheyne by every nighte at ten of the clocke to be locked crosse the great bridge and soe kept untyl five of the clocke in the morninge.'

Barford Bridge

Below Bedford no road-bridge crosses the Ouse for a distance of about six miles, but near Great Barford Church the road to Blunham crosses by a mediæval bridge (Fig. 48) having seventeen pointed arches, spanning a distance of over 150 yards. The width between parapets is 16 feet, the bridge having been widened upstream in brick. It is a pity that the builder was not content with using an ordinary type of arch-ring, as the missing-teeth or cog-wheel variety is hardly pleasing.

In his will, which was proved in 1429, Sir Gerard Braybroke expressed the wish that 'the bridge of Berford in Beds. be performed and finished' with his goods, and in the year 1446 the burgesses of Bedford appealed that the fee-farm rent might be remitted, giving as one reason 'the building of a new bridge over the Ouse at Great Barford.' From this it appears that a bridge was certainly built here in the middle of the fifteenth century, probably the one at present in use.

In 1753 it was presented to the Sessions that Barford Bridge ought to be repaired by the County, and in 1777 an account was received from Stephen Hart for work done on the bridge, while in 1791 a bill for bricklayers' work was recorded. Eight years later the inhabitants of the County of Bedford were indicted for not repairing 'a common bridge over the River Ouze called Barford Bridge, which was in decay.' In

49. ST. NEOTS BRIDGE.

50. NUNS' BRIDGE.

51. ALCONBURY BRIDGE.

52. SPALDWICK BRIDGE.

February 1814 the sum of £2 5s. 6d. was paid for 'clearing ice away from Barford Bridge.'

Repairs to the bridge cost £2 10s. 4d. in 1815, 13s. 0d. in 1818, and many additional bills were presented the following year. In 1824 damage caused by flood cost the sum of £39 13s. 7d. A special drawing of the bridge was made for the Sessions in 1825.

Two and a half miles below Great Barford the *River Ivel* Ouse is joined by the river Ivel, which, rising near Baldock, flows north through Stotfold to Astwick. Shortly below Astwick it is joined by the river Hiz, and a mile north of Langford by a branch from Clophill and Shefford.

Considerable work was done on the bridges at Shefford in the year 1803, four water piers of the North Bridge and three of the South Bridge being rebuilt. In spite of this, both were said to be 'in decay' the following year. The rebuilding of these bridges was under consideration in the year 1828. Both are now iron bridges.

Langford Bridge was severely damaged by flood in the year 1817, and it was reported to the Sessions that 'one fourth in length of the centre pier has fallen down and with it about the same quantity of the bridge, which is reduced to about 9 feet in width and is in a dangerous state.' The bridge was described as 'an ancient structure, consisting of two Gothic arches supported on a centre pier $4\frac{1}{2}$ feet wide which blocks up a great part of the river Ivel.' It was recommended that a wooden bridge be built with brick abutments, as a single-span arch would raise the road too high. The following year the decision was taken to rebuild with an iron arch, and the old bridge was pulled down. The new bridge was completed in September 1819, and is probably the one now in use. The bridge at the northern end of Biggleswade, carrying the road to Sandy, bears the date 1758. The three segmental arches span a distance of 16 yards, and each has triple keystones projecting beyond the arch-rings. Bishop Dalderby in 1302 granted indulgences to those contributing towards the bridge of Biggleswade, and in July 1372 pontage was granted for a period of three years. It was shown by Ogilby as a stone bridge.

Biggleswade Bridge

Owing to complaints concerning the state of Biggleswade Bridge, John Millington in 1818 was asked to make a report. According to him, the 'dilapidations were trifling, but the foundation of one of the wing walls was undermined to a slight degree by the wash of the water when cattle and carriages went into the watering place.' He advised the underpinning of the wall, which should not cost more than £6.

Wash Brook Bridge, adjoining Biggleswade Bridge, a timber structure, was reported in 1821 to be in a very decayed state. It was reconstructed two years later. Girtford Bridge at Sandy is of the same type as Biggleswade Bridge.

At Sutton, about three miles north-east of Biggleswade, an interesting packhorse bridge, across a tributary of the Ivel, can still be seen. It has two pointed arches said to date from the fourteenth century.

The last bridge over the Ivel before it joins the Ouse carries the road from Blunham to Tempsford, and is called Blunham Bridges. It has five semicircular arches with chamfered arch-rings, features common in bridges built late in the seventeenth century. The width between the brick parapets is 13 feet.

It is interesting to find from John Armstrong's *Survey of the Great Post-Roads between London and Edinburgh*, published in 1776, that even at such a comparatively recent date the only crossing over the Ouse at Tempsford was by ferry. A bridge must have been built soon after that date, as in October 1820 it was certified to the Sessions that 'Tempsford Bridge, which has been built by the inhabitants of the County of Bedford, in place of the old one of that name, is now open for horse and foot passengers.' Three years later John Millington, the County Surveyor, made a long and complicated report on work to be done to make good damage caused by farm carts passing through the arches of the bridge, and a further report was drawn up in July 1824. The closing of the arches appears to have hampered the farmer of the adjacent land, as, in April 1828, William Saunderson of Roxton made a formal application for permission to pass through one of the arches with his hay carts, and to remove one of the posts which obstructed his so doing. *Tempsford Bridge*

The first mention of the bridge of St. Neots is a tale recorded by Matthew Paris that William, Earl of Ferrers, who from his youth had suffered from gout and had to be drawn from place to place in a chariot, was, in April 1254, thrown from the bridge of St. Neots through the carelessness of his driver, and died of his injuries. Pontage was granted in July 1388 for the repair of 'Seint Neet brigge,' and in 1439 Roger Benethon left the sum of 3s. 4d. 'to the building of *St. Neots Bridge* the bridge of St. Neots.' In his *History of Eynesbury and St. Neots*, Cornelius Gorham quotes a Harleian MSS. in the British Museum from which it appears that the present stone bridge was not in existence in the year 1588. At an inquisition taken that year it was stated that 'the long Bridge or Causye conteyneth in Length 704 feet, whereof 43 Arches wholly built of timber conteyne in Length 448 feet, Breadth $10\frac{1}{2}$ feet. Also 29 Arches of tymber work of severall heighthes and lengths of 7, 9, and 12 feet apeece, built uppon a stone wall of 6 feet high, Length 256, breadth $7\frac{1}{2}$ feet.'

The present stone bridge is thought to date from early in the seventeenth century, as mention is made in the Calendar of State Papers (Domestic) for the years 1616 and 1617 of large sums of money paid for work on the bridge. Accounts for its repairs appear in the Sessions Rolls for 1752, and again in 1756 and 1758. Those for the year 1789 included such items as stone, brick, iron and lead work. In 1799 the bridge was said to be ruinous, but the repairs were completed by February 1801.

After this date no further trouble is recorded for twenty-three years, but in March 1824 a very full report was submitted by Mr. C. Bevan dealing with the question of floods at St. Neots. In time of flood the river flowed over the road at Eaton Ford, and he advised the construction of four new tunnels, 12 feet in diameter, through the causeway. He also considered the question of widening the bridge, and estimated that the cost of making a projecting footpath $3\frac{1}{2}$ feet wide on each side would be about £700. If the whole bridge was widened 10 feet on the south side by three cast-iron arches, he thought that the cost would not exceed £200. The possibility of the flooding of the river was still causing anxiety during the following year, but there is nothing to show that the recommendations were carried out. The repairs to the bridge in 1827 amounted to £17 6s. 5d.

St. Neots Bridge, as it stands to-day (Fig. 49), has three segmental arches over the main stream, spanning about 50 yards. Two of these have chamfered ribs, connected by cross ribs, a very unusual feature. Farther west there are several semicircular arches with flat pedestal buttresses, and on one is a panel with the date 1647. The scheme for building out the footways, as suggested in 1824, was evidently adopted, as these are now carried by a series of small segmental arches corbelled out from the sides of the bridge. The width between the parapets is now about 22 feet.

A mile below St. Neots the Ouse is joined by the river Kym, but none of its bridges are of special interest. Ogilby showed a stone bridge at Staughton Highway, but the present one is built of brick and concrete. The timber bridge he showed on the site of Hail Bridge has been replaced by one of brick and stone. *River Kym*

A stream called the Alconbury Brook joins the Ouse about a mile above Huntingdon Bridge. Nuns' Bridge (Fig. 50), which carries the road to Brampton, is its best bridge, as it still has three arches probably dating from the fifteenth century. Two of these are pointed, the third being segmental. They have double arch-rings, built in two orders, both being chamfered. This part of the bridge has been widened *Alconbury Brook* in brick by about 11 feet, and the three southern arches have been entirely rebuilt.

Another bridge (Fig. 51) with pointed arches, but of later date, crosses the river in the village of Alconbury. It has, however, been much repaired and partly rebuilt in brick. The width between its brick parapets is about 10 feet.

Spaldwick Bridge (Fig. 52), over the southern branch of the Alconbury Brook, carries the road from Huntingdon to Thrapston. It also has pointed arches, three in number, the middle one having chamfered ribs. The parapets are built of brick and allow a roadway of nearly 24 feet, the bridge having been widened downstream by about 13 feet.

Unfortunately nothing appears to be known of the history of these three attractive bridges, nor of Hammerton Bridge, about four miles north-west of Alconbury, which has ancient stone piers carrying a modern timber deck.

Huntingdon Bridge

A short distance before reaching Huntingdon Bridge, the North Road from Royston crosses a long causeway known as Godmanchester Bridge. It was reconstructed in 1637 by Robert Cooke, but considerable repairs were done a hundred and thirty years later, and it now consists of two bridges each with eight segmental arches.

Huntingdon Bridge (see frontispiece) is undoubtedly the finest mediæval bridge in this part of England, and although it has often been stated that the present bridge was built about the year 1300, the evidence obtained from the Patent Rolls and other contemporary documents leads to the conclusion that the building took place nearer the end of the fourteenth century. The arch nearest the town, however, probably belonged to the earlier bridge. It is considerably lower than the others and widened in a very unusual manner by splaying the vault outwards on each side, evidently to make this arch more in keeping with the larger ones then being built. The arch-rings are formed by ribs finished with hollow mouldings, and the drip-course built over this widened arch is continued round both faces of the adjacent cut-water.

Five of the six arches of the bridge are pointed in shape with triple arch-rings built in three orders. They are not chamfered, however, except for a short distance at the haunches of the inner arch-ring. The total span is about 80 yards, and the width between parapets 17 feet, but recesses at road level are provided over each of the very massive cut-waters. The upstream parapet over the second and third arches nearest the town is carried by an attractive trefoil corbel table.

A bridge existed here early in the thirteenth century, as in a Quo Warranto plea of 1259 it was stated that Huntingdon Bridge is 'shaken and impaired by the carriage of dung and corn,' and orders were given that the Sheriff should 'bind all who carry dung beyond the bridge to contribute for the support thereof.'

A commission to Gilbert de Preston is recorded in the Patent Rolls for January 31st, 1265, instructing him to find by inquisition 'what rents are due for the repair of Huntingdon Bridge and to determine who are bound to repair it.' A further inquisition was taken in 1276, and it was then stated that the bridge was 'so broken that it is almost impassable for passengers on horseback or on foot: ... and that they cannot convey over it their implements of husbandry etc.' It was ordered that the bridge 'should be repaired by the inhabitants of the whole county, and they shall make reparacion within 15 days of the feast of St. Martinmas under penalty of forfeiting £100.'

Pontage for a period of three years was granted to the bailiffs and good men of Huntingdon on November 6th, 1279, and the tolls included 'on every Jew or Jewess crossing the bridge on horseback one penny, on foot one half-penny.' Jordan de Houcton was appointed Warden of the work, with five others to assist him.

A severe flood laden with ice destroyed the bridge during the winter 1293-4, and collections were made in churches during 1296 for its repair. According to the Close Rolls for April 1300, instructions were sent to Hugh de Despenser, justice of the Forest of Wauberge, ' to provide 24 oaks fit for timber, where this can be done with least damage to the King, for the works of the bridge of Huntingdon, of the King's gift.' This might be taken to imply that it was then a wooden bridge, but was not necessarily so, as timber was often required for piles or for the protection of the piers of a stone bridge.

The Patent Rolls record for April 1329 that Huntingdon Bridge 'is in a ruinous state, in many places broken through, and at one part threatening to fall, and that divers legacies, bequeathed in the past for its support, for the want of a keeper of the same, have not yet been used for the purpose.' Philip de

Ravele, chaplain, was then appointed by the King as 'Keeper of the Bridge,' and it was ordered that any surplus funds were to be used to build 'a chapel on the bridge in honour of St. Thomas the Martyr and St. Katherine the Virgin.' Evidently this chapel was duly built, as reference is made in the Rolls of Parliament for 1334 to 'a little chapel lately built on the bridge of Huntingdon,' when Robert de Fenere, parson of the Church of St. Clements, prayed 'to have the keeping of the said chapel annexed to his church together with the charge of the bridge.' According to the Historical Monuments Commission, the chapel was dedicated to St. Thomas of Canterbury.

In April 1344 a grant of pontage was issued to the Warden of the Bridge of Huntingdon for a space of five years, and in March 1356 pontage for its repair was granted to the burgesses, on this occasion for only three years.

The Close Rolls for the year 1364 contain a further reference to this bridge, recording that 'Robert de Thorpe and his fellows' were appointed to survey Huntingdon Bridge, 'which has been long ruinous.' They stated that 'it had been found in the time of Edward I that the commons of the whole county are bound to repair the said bridge at their own cost.'

In 1370 Robert de Thorpe and his fellow-justices were appointed to survey the bridge, which was stated to have 'long been broken and ruinous.' Orders were given 'to open all the water gates of mill-ponds upon the said bridge whereby the water is held up, and to keep them open until the foundations of the said bridge shall be built as needful is, and further, by amercements and other lawful means as they shall see fit, order that the recovery or new building thereof be not delayed or hindered.' It was stated that 'the bridge is flooded up to the top by the water of the said ponds and owing to the obstruction thereof.'

From this it appears that extensive rebuilding took place about the year 1370, especially as seven years later, according to the Rolls of Parliament, the Countess of Norfolk complained that, contrary to her franchise, her tenants had been compelled to contribute to the building of the bridge at Huntingdon.

St. Ives Bridge

At St. Ives, about five miles below Huntingdon, another ancient bridge (Fig. 53) crosses the Ouse. Four of the arches are pointed in shape, each having five ribs, and probably date from early in the fifteenth century. The other two arches were rebuilt in the year 1716.

On the central pier stands the Chapel of St. Leger, which was until recently surmounted by a brick dwelling-house bearing the date 1736. Owing, however, to the weight of this superstructure, the walls of the chapel were found to be giving way, and in 1929 the brick building was removed, leaving the chapel in its original form. The total span of the six arches is 67 yards, and the roadway is 13 feet in width.

Mr. C. H. Evelyn-White, F.S.A., in vol. i. of the *Transactions of the Cambridgeshire and Huntingdonshire Archæological Society*, gives an excellent drawing and description of the bridge and chapel. He also refers to an extract from a Roll of 1259,

53. ST. IVES BRIDGE (before the restoration of the chapel).

54. LITTLE CHESTERFORD BRIDGE.

55. ABBOT'S BRIDGE, BURY ST. EDMUNDS.

56. MOULTON BRIDGE.

which specifically mentioned the bridge of St. Ives. According to the *Victoria County History*, it was a wooden bridge until the year 1384, and the chapel was consecrated in the year 1426.

Earith Bridge
The bridge of 'Ereth' was mentioned in a perambulation made in the year 1286, according to the Cartulary of Ramsey Abbey, and in September 1346 a commission was appointed to determine who was responsible for its maintenance, it being reported as 'now removed by default of repair.' The present bridge is of the suspension type.

Three miles below Earith the old road from Ely crossed the Ouse by Aldreth Causeway and High Bridge, which since early times formed the entrance to the Isle of Ely. According to Mr. C. H. Evelyn-White, F.S.A., King Stephen, crossing here, by means of a bridge of boats and hurdles, became master of the Isle. Like so many bridges, it continually suffered from lack of attention, and finally ceased to exist not many years ago.

River Cam
The river Cam, the largest tributary of the Ouse, consists of two rivers, the Granta and the Rhee, which join a few miles above Cambridge. The former rises near Widdington in Essex and flows north past Saffron Walden, but the source of the Rhee is in Bedfordshire, about five miles west of Royston.

Above Cambridge few of the bridges are of interest. Several across the Rhee have stone keystones and may date from late in the eighteenth century, and only two over the Granta call for remark—one about a mile west of Saffron Walden, and the other a few miles farther north, at Little Chesterford. The former bridge, which is situated close to Audley End House, was built by Robert Adam in 1771. The three arches are semicircular, with elaborately moulded arch-rings, spanning a distance of 25 yards, with a width between the balustrades of nearly 20 feet.

Little Chesterford Bridge (Fig. 54) is a far humbler structure, with two brick arches and parapets bearing the inscription 'G X K 1791.' Its middle pier and also the abutments, being built of stone, appear to belong to an earlier bridge. In 1770 it was described by Peter Muilman as being ' mostly of stone.'

At Whitlesford, about seven miles south of Cambridge, the Granta was crossed by a bridge as early as 1278, according to Mr. C. H. Cooper in his *Annals of Cambridge*. In September 1637 the Corporation of Cambridge ordered that 'the farmer (of the tolls) of Whittlesford Bridge should thenceforward bring the mace, whereby he gathered toll, to the Hall on Michaelmas Day yearly and then tender the same to the Mayor.'

Bridges at Cambridge

The name Cambridge does not appear to have been used before the beginning of the fifteenth century. Its earlier form was Grantabrige (Grantabrycge in the Anglo-Saxon Chronicles of 875), changed about the middle of the twelfth century to Cantebrig, and the river still bears the name Granta or Cam.

Many historical notes concerning the bridges of Cambridge are given by Thomas D. Atkinson in his book entitled *Cambridge Described and Illustrated*, and the *Annals of Cambridge*, by Charles H. Cooper, Town Clerk of Cambridge in the middle of the past century, records all important events concerning

Cambridge from the year 1276 until well into the nineteenth century.

The most important bridge was the one near Magdalene College, formerly known as the Great Bridge, which was commanded by the Castle. The bridges over the two streams which carry the present Silver Street bore the name of the Small Bridges. All these were originally in the hands of the King, and the Sheriff maintained the Great Bridge out of charges on certain lands. During the reign of Edward I the Great Bridge was in a ruinous state, and a ferry was provided by the Sheriff, who appropriated the income to his own use.

A commission in the year 1278 stated that the bridge was in 'scandalous decay,' but there is no record of any work being done to improve matters, and in March 1349 a further commission was directed to enquire who was responsible for the repairs. The bridge was then said to be in a dangerous state, but it appears that no conclusion was reached, as six years later another commission was appointed, and a further one in May 1362. Pontage was evidently granted, as three years later an audit of the accounts was ordered. In 1390 the bridge was reported to be 'dilapidated,' and it continued in this condition for many years. In June 1423 it was stated that 'people were said to have removed stones and timber from the bridge,' and in June 1488 a commission was appointed to enquire by a jury of the County who ought to repair the bridge, and to 'distrain and compel those, who are so bound, to repair it with all speed.' Lysons stated that the bridge was rebuilt in 1482, but this seems doubtful, as the fact is not recorded in the *Annals of Cambridge*. It evidently took place, however, early in the next century, as in 1499 the taxing of land was increased to £3 per hide for 'new building the Great Bridge.' This was decreased to 13s. 4d. per hide in 1546, an amount considered sufficient for 'the maintenance of the bridge.'

This bridge, however, survived less than a hundred years, being carried away by flood in September 1594, as were St. John's and King's College Bridges. During the Civil War all bridges except the Great Bridge were pulled down, but this bridge was guarded by a cannon called 'a Drake.' The accounts for 1643 include under the date of March 19th an item: 'Payed to Soloman Little for watching the Bridge, the Posts and Rayles being pulled down and the Rayles ready to fall into the water—1s., other expenses, including removing the drake—18s. 11d.'

The Great Bridge was rebuilt in 1754, but in 1799 it was again reported as ruinous, and the present iron structure was built in 1823 to the design of Mr. Arthur Brown. The bridge constructed in 1754 was of stone, but all the earlier ones were built of timber, so that the bridge, with six semicircular arches, depicted on the first Common Seal of Cambridge made in 1423 and the one, with four pointed arches, on the Mayor's Seal used in 1471 were imaginative.

The Small Bridges

The Small Bridges were also of timber until the present ones were built in 1841. Like many mediæval bridges, these had a chapel attached, and 'a licence to celebrate divine service in Small Bridges Chapel in the suburbs of Cambridge' was issued late in the fourteenth century. Indulgences for the repair of these bridges were also granted.

57. BRANDON BRIDGE.

58. HOUGHTON ST. GILES (FOOT) BRIDGE.

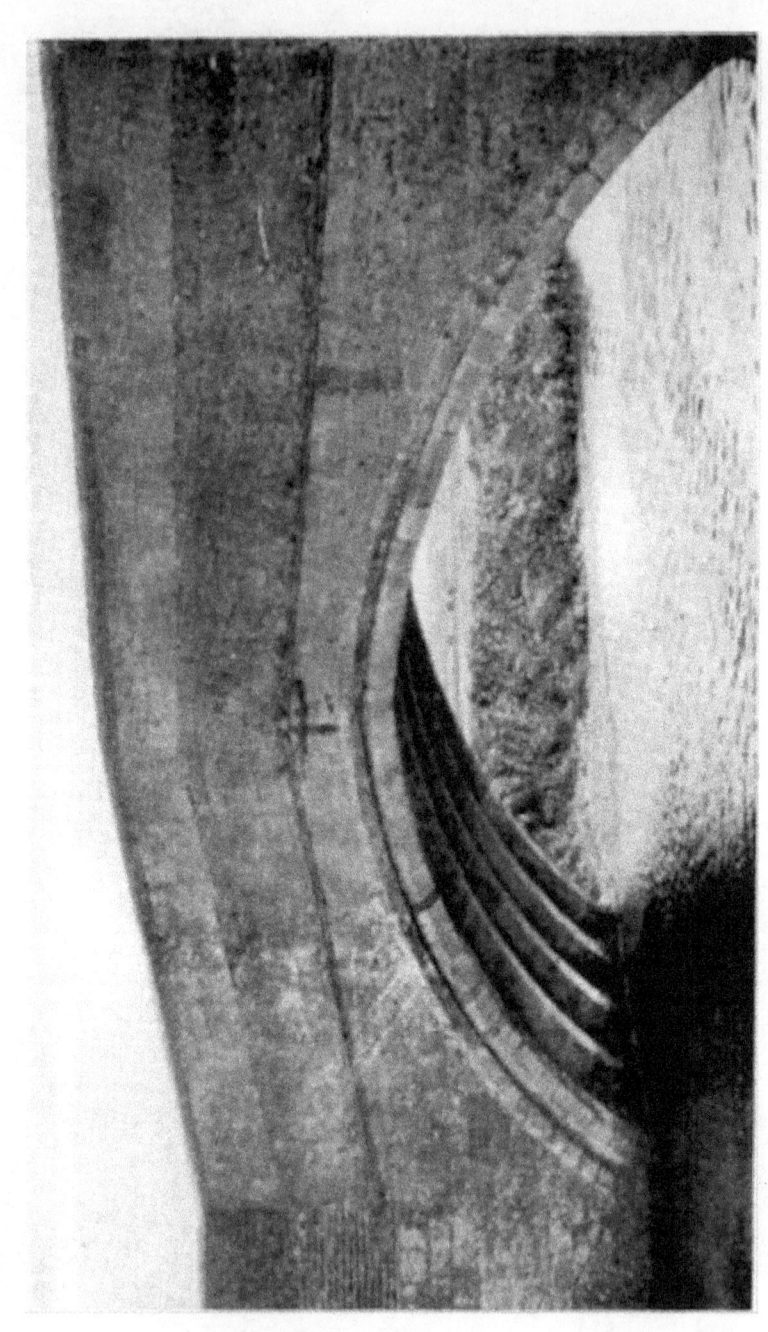

59. Wiveton Bridge.

In October 1399 pontage was allowed for a period of two years to John Jaye the 'heremite' for the repair of the bridges and causeway between Cambridge and Barton. In a further grant, made in June 1407 to Thomas Kendale, the name 'Smalebrigges' is actually used. These bridges were destroyed during the Civil War in 1648, and again in 1773, and were rebuilt by the Corporation. The present 'elegant iron' arch was constructed in the year 1841.

Garrett Hostel Bridge

A bridge near the Garrett Hostel was washed away in November 1520, and at the Assizes held in March 1627 the 'inhabitants of Cambridge' were indicted for the non-repair of the bridge. Shortly afterwards it was repaired by the Corporation, only to be demolished during the Civil War. It was rebuilt with the help of a 'free contribution from Trinity College and Trinity Hall' in 1646, and again in 1769 and 1822, each time in timber. The iron bridge was constructed in 1837.

King's College and Clare Bridges

Thomas D. Atkinson gives an excellent map of Cambridge on which King's College Bridge is given the date of 1818. He gives Clare Bridge as dating from the year 1640, and it is certainly the only ancient bridge now in Cambridge. The three segmental arches have a total span of 27 yards and the width between the balustrades measures $14\frac{1}{2}$ feet. It is certainly one of the most attractive seventeenth-century bridges remaining in England.

St. John's Bridge, constructed from Wren's design between the years 1696 and 1712, has three segmental arches spanning a width of 24 yards. The balustrades are slightly over 13 feet apart. The Bridge of Sighs was built in 1826.

J. Moor in his map of the Fens published in 1684 shows few bridges north of Cambridge. The last shown over the Cam was the one at Clayhythe, but over the Ouse itself he shows a bridge a mile southeast of Ely and also one at Littleport. All the present bridges on these sites are modern, as are those between Littleport and the Wash.

CHAPTER VII

THE RIVERS AND BRIDGES OF EAST ANGLIA AND ESSEX

THE RIVER LARK

RISING about three miles south of Bury St. Edmunds, the river Lark and its tributary the Kennett drain the district lying east and north of Newmarket.

Bridges in Bury St. Edmunds

Until about the year 1840 a mediæval bridge in Bury St. Edmunds, known as Eastgate Bridge, carried the road leading to Diss. It has unfortunately been destroyed, but a sketch of the bridge, showing five pointed arches, is given in vol. iv. of the *Proceedings of the Suffolk Institute of Archæology*.

Close to the site of this bridge, the wall of the abbey precincts is carried over the river Lark by three pointed arches (Fig. 55) which, in addition to carrying the wall, also support a foot-bridge about 5 feet in width inside the precincts. The openings in the buttresses, built against the outer face of the wall, suggest that there was once a plank bridge on this side. The arches carrying the inner bridge, however, are almost segmental in shape, but both the pointed and the segmental arches have chamfered ribs. This bridge, known as the Abbot's Bridge, is said, on the authority of Grosse, to date from about the year 1230, but the shape of the arches indicates that it was probably built at a later date, possibly after one of the serious conflicts between the abbey

ANCIENT BRIDGES

and the town. These occurred in the year 1264 and again in 1327, when the townsmen attacked the monastery and the great gate of the abbey was burnt down. A stone bridge near the east gate was mentioned in the Town Rental for the year 1295.

Most of the bridges below Bury St. Edmunds have been rebuilt during the nineteenth century, but at Icklingham the road to Caversham is still carried by a timber bridge on brick piers. The last crossing of the Lark, at Prickwillow, before it joins the Ouse is a girder bridge built in 1866.

River Kennett

The river Kennett, however, is crossed at Moulton by a very attractive packhorse bridge (Fig. 56) having pointed arches said to date from the fifteenth century. The arch-rings are of brick, but the rest of the bridge is built of flint and stone. The four arches span a distance of 20 yards and the width between parapets is 5 feet. About two miles north of this bridge, alongside the modern concrete bridge and ford at Kentford, are the remains of another ancient bridge built also of flint and brick. Only parts of the piers now remain, 6 feet in overall width, but there are signs of a central rib in the springings of one arch. The present concrete bridge took the place of a brick one having three arches. Ogilby did not show any bridge at Kentford, but made the remark, 'Water running in the Way.'

THE LITTLE OUSE

The Little Ouse rises in Redgrave Fen, which is also the source of the river Waveney, and forms for many miles the boundary between Norfolk and Suffolk. The only bridge of note before reaching

Thetford is one at Rushford, which has a single pointed arch of about 4 yards' span. The arch-rings are of stone, but the rest of the bridge is faced with flints and appears to have been recently restored. The roadway is $13\frac{1}{2}$ feet in width.

Three miles below Rushford the Little Ouse receives a tributary from the south which, in the village of Ixworth, is crossed by a pleasant brick bridge with five semicircular arches, erected probably early in the nineteenth century. This was evidently not the first bridge at Ixworth, as Hewe Baker the elder, in his will of 1545, left the sum of 40s. 'to the making of one bridge of lyme and stone at the mylle of Ixworth.'

Sapiston Bridge over this stream, between Sapiston and Honington, has the date 1780 carved on a stone on one of the brick parapets, and the bridge at Fakenham bears the date 1848. Each of these bridges has three brick arches.

The last bridge over this stream, carrying the road from Euston to Barnham, has two segmental stone arches with projecting keystones, probably built towards the end of the eighteenth century.

A further tributary, the river Thet, joins the Ouse at Thetford. All its bridges are, however, comparatively modern, except the one on the outskirts of Thetford called Melford Bridge. This is built of brick with stone arch-rings and bears, on the upstream face of its middle pier, a stone on which is carved a coat of arms and an inscription stating that the bridge was built by John Woodhouse in the year 1665. The two arches have a total span of 7 yards and the roadway is slightly over 12 feet in width.

River Thet

In 1739, according to Blomefield, the ducking stool of Thetford was kept alongside the Nuns' Bridge, and the 'Nones briges' were mentioned in an indenture of 1539 concerning the 'makynge, kepynge and reparynge of the said briges.' The present Nuns' Bridge has brick arches.

During the reign of Charles II an Act was passed to make the river navigable to Thetford. The earlier limit was the 'White House' near Brandon Ferry.

Brandon Bridge

A cast-iron bridge, built in 1829, crosses the Little Ouse in Thetford, but at Brandon, about seven miles below, there is a massive masonry structure having four arches spanning a width of 30 yards. Here again the arch-rings are built of stone, while the rest of the structure is of brick. As will be seen from the illustration (Fig. 57), this combination of materials has proved very elastic, one arch having crumpled up in a most extraordinary fashion, but remaining perfectly stable for many years. On each side of the bridge are large cut-waters, two having recesses at the road level. The width between the parapets is 12 feet.

According to the Calendar of Inquisitions, the tolls of 'the bridges of Thetford, the ferry at Santon, and the bridge at Brandon ferye' were held in 1326 by Ralph de Cobham, and in July 1330 pontage for a period of five years was granted to the Bishop of Ely for 'the repair of the bridge at Brandonferry.'

Thomas Martin, in his *History of Thetford*, stated that in the reign of Queen Elizabeth the tolls of 'the Bridge of Incelland or Selford in Norfolk and those of Brandonferry, Honiton and Euston bridges shall be held in fee farm of the Queen at a rent of

£8 6s. 8d., but the corporation shall keep them in good repair.' From the Corporation records it appears that in the time of King Edward VI the Brandon Bridge tolls were leased at £4 a year.

Another tributary of the Ouse, the river Wissey, is received about three miles south of Downham Market. It rises near Shipdham, four miles southwest of East Dereham, and had, according to Faden's map, no bridges above Hilborough at the end of the eighteenth century. There were, however, several over the branch from Carbrooke, but all appear to have been rebuilt during the past century.

River Wissey

Below Hilborough there are several brick bridges with stone keystones, dating probably from early in the nineteenth century, and the one near Mundford, which has a pointed brick arch, is certainly not ancient. In Ogilby's days there was a stone bridge on this site.

An iron bridge, built in 1899, now crosses the river at Stoke Ferry, and it is probable that earlier ones were built of timber. Stoke Ferry Bridge was the subject of an enquiry in the year 1291, at which it was stated that 'between the piles in the middle of the bridge at Stoke Ferie there ought to be a space of 16 feet, now narrowed by 7 feet in breadth.' The King's Justice found that the bridge was broken and ordered the Sheriff to distrain the hundreds of Chakelose and Gumeshogh, but the jurors expressed a doubt as to 'whether of right they ought to do so because it was first built by alms.'

The last bridge over the Wissey is at Hilgay, where Ogilby showed a timber bridge carrying the road from Ely to Downham Market.

River Nar   The Nar joins the Ouse at King's Lynn, where, in the fourteenth century, the bridge across it was called 'Suthbrigge.' It is recorded that in the year 1364 'part of it called the Draught, next the borough, and made by the men of the said borough, was in good repair,' but the remainder was in a bad state and no one was responsible. A map of King's Lynn made in 1725, and published in Blomefield's *History of Norfolk*, gives a list of twelve bridges in the town. They are mostly over small channels, tributaries of the Nar and the Gaywood River.

The Nar is now crossed by girder bridges at Castle Acre, West Acre, and Narborough, but the bridges at East and West Lexham are built of brick.

THE BABINGLEY AND HEACHAM RIVERS

Both these streams flow into the Wash, the former only a few miles north of the mouth of the Ouse. None of their bridges are ancient, although several are built in the mediæval style; one over the Babingley, near Flitcham, has embattled parapets, and both bridges in Heacham, across the Heacham River, have pointed arches. Another bridge, which carries the road from Hunstanton to Burnham Market, has a chamfered rib on each side over a pointed arch.

THE RIVERS BURN, STIFFKEY, AND GLAVEN

These three rivers drain the northern parts of the county of Norfolk. They are small streams, crossed by numerous bridges, mostly built of brick. None over the Burn are of interest, but half a mile northeast of Houghton St. Giles the river Stiffkey is crossed by an attractive foot-bridge (Fig. 58) having the date 1709 carved on one of its keystones. The single arch is built of brick, but the keystones are of stone, as are the subsidiary keystones placed near the haunches of the arch, which form a unique feature. Iron railings take the place of the usual parapets, and the overall width is about 6 feet.

The stone bridge over the river Glaven at Wiveton (Fig. 59) is of more than ordinary interest, as it has a pointed arch with five chamfered ribs, and is the only one of this type in the district. The total span is 12 yards and the width between the parapets 12 feet. Unfortunately nothing appears to be known of the history of this bridge, the only reference in mediæval documents being a bequest in the will of Robert Paston of Wiveton, made in the year 1482, of sixpence to the ' repacon capelle Ste Trenit sup ponte,' which may possibly refer to the bridge, although no chapel now exists.

Wiveton Bridge

THE RIVER BURE

The river Bure rises a few miles west of Saxthorpe and flows in a south-easterly direction for the greater part of its course before joining the river Yare in Great Yarmouth. There are now concrete or iron girder bridges at eight of its crossings, but brick bridges still remain at Aylesham, Buxton, and Hoveton St. John. None of these, however, are ancient, and the stone in the parapet of the bridge at Hoveton St. John bearing the date 1614 probably belonged to an earlier bridge.

Mayton Bridge

Two miles north-west of Coltishall the road from Little Hautbois to Hainford crosses the old channel of the Bure by a bridge shown by Faden in 1797 as 'Maiden Bridge.' Blomefield refers to it as Mayton Bridge, which is probably the correct name, as it is situated less than a quarter of a mile from Mayton Hall. This bridge (Fig. 60) is particularly interesting, as its two arches are four-centred in shape, a very unusual feature in brick bridges. Nothing appears to be known of its history, but it probably dates from the fifteenth or sixteenth century. The little shelters built in brick at each end of the upstream parapet are also unique. The total span is about 9 yards and the parapets are slightly less than 11 feet apart. The main stream is crossed by a timber bridge on brick piers.

Wey Bridge

The only bridge across the Bure that appears to be noted in mediæval documents is the one, known as Wey Bridge, situated about a mile north-east of Acle and about eight miles west of Great Yarmouth. It was mentioned in 1346 in a charter to the Abbot and Convent of St. Benet, Hulme, concerning the fishing in the river, and pontage for its repair was granted in July 1408. The present bridge (Fig. 61), spanning about 25 yards, has three semicircular arches, built of brick and stone. They have triple arch-rings built in three orders, and the roadway is 13 feet in width.

Heigham Bridge

Three miles above Wey Bridge the Bure is joined by the river Thurne, the outflow of Hickling Broad and Heigham Sound. A mile south-east of Potter Heigham the road to Rollesby crosses the Thurne by an interesting stone bridge (Fig. 62), called Heigham

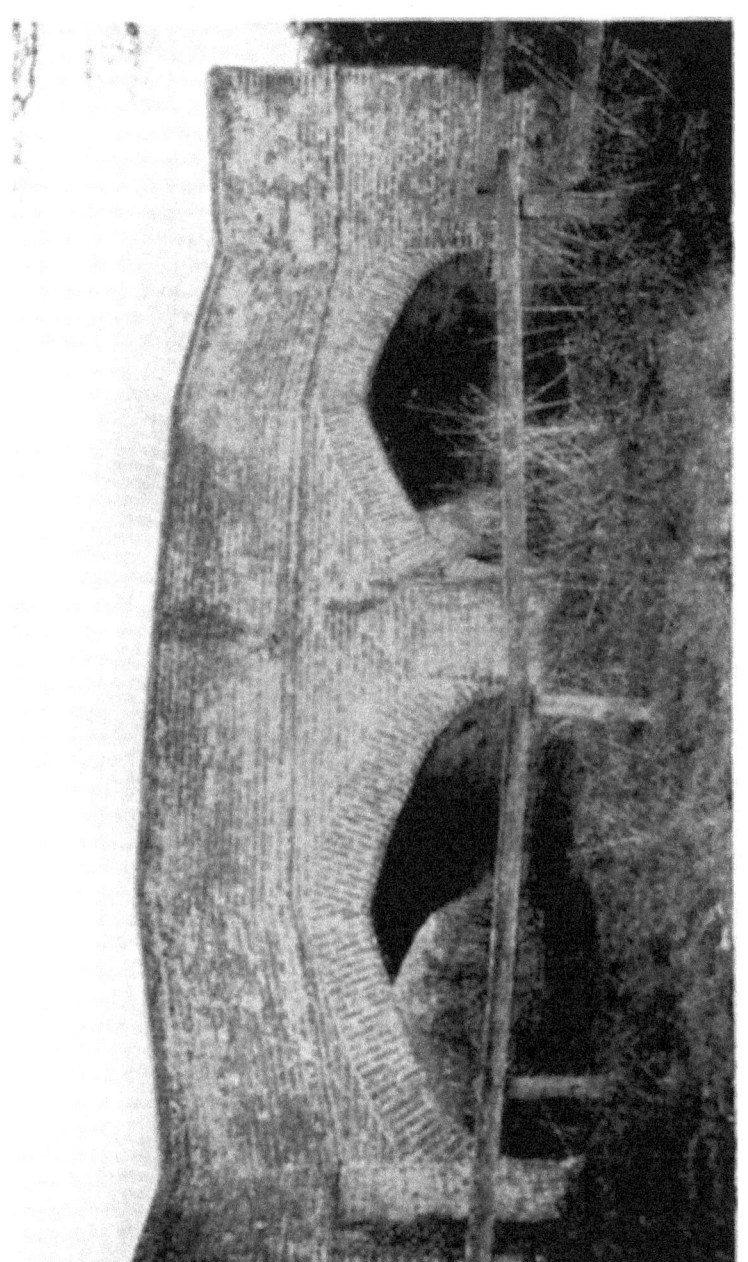

60. Mayton Bridge.

61. Wey Bridge.

62. Heigham Bridge.

Bridge, which has one segmental and two pointed arches, each of the latter having four widely chamfered ribs. The segmental arch is of later date. The total span is about 19 yards and the parapets, which are built of brick, are 12 feet apart.

THE RIVER YARE

Rising a few miles south of East Dereham, the Yare, and its chief tributary the Wensum, drain the whole of Central Norfolk. Above Bickerston Bridge, which has a single brick arch, there were, according to Faden, only fords at the end of the eighteenth century, and it is doubtful if any of the existing bridges above Cringleford are more than a hundred years old.

Cringleford Bridge, however, probably dates from early in the sixteenth century, as Blomefield records that 'in 1519, on St. Leonard's Day, happened a Flood, which overflowed great Part of the City (Norwich), and broke down Cringleford Bridge.' This bridge (Fig. 63), which has two four-centred arches, spanning a distance of 18 yards, is built of stone, but has been repaired and widened with brick, and both sides of the bridge are faced with stone. The width between parapets is 15 feet. *Cringleford Bridge*

Harford Bridges, about two miles below Cringleford, on the road from Norwich to Ipswich, have elliptical brick arches on stone abutments, widened in brick to give a roadway of over 20 feet. Ogilby towards the end of the seventeenth century gave the name as Hartford Bridge, and Blomefield in 1736 wrote: 'I rather think it took its name from the Family sir-named de Hereford, one of which first *Harford Bridges* built a Bridge here in King John's Time.' 'Hertford brigge' was, in the reign of Henry VIII, one of the boundary points of the City of Norwich.

River Tas A mile and a half below Harford Bridges the Yare is joined by the river Tas, which rises about twelve miles south of Norwich. In Newton Flotman the river is crossed by a bridge (Fig. 64), carrying the road from Norwich to Ipswich, which is partly mediæval, as three of its arches are of stone, each with four chamfered ribs. The other two arches are four-centred, one being built of stone and the other, of much later date, of brick. This type of arch has also been used for the widening on the downstream side. The total span is 34 yards, and the roadway is now 24 feet wide, compared with its original width of about 8 feet. The upstream parapet bears the date 1838. In a book entitled *Excursions through Norfolk* it is described as 'a good brick arched bridge,' so that the first widening must have taken place before 1818, the year in which the book was published.

Trowse Bridge A girder bridge now spans the Yare at Old Lakenham, and there is another of this type over the southern stream of this river at Trowse Newton. The bridge over the northern branch, which bears the date 1863, was widened in 1913. 'Trows' Bridge was mentioned as early as the year 1430 and 'Trowys' Bridge was one of the boundaries of the City of Norwich in the reign of Edward VI. According to Blomefield, 'Trowse Bridge, the utmost limit of the city,' was the place where Charles II and his Queen were received on their visit to Norwich in September 1671.

A short distance below Trowse Bridge the Yare receives its largest tributary, the river Wensum, which rises at West Rudham, about five miles west of Fakenham. The Wensum flows through Great Ryburgh, Billingford, and Lyng to Lenwade, but there are no ancient bridges on this part of the river. Lenwade Bridge, shown by Ogilby in 1674 as 'Leanard Bridge,' and mentioned six years earlier by Philip Skippon as 'Leonard Bridge,' is now a brick structure with three semicircular arches, with a roadway 13 feet in width. A new concrete bridge, built alongside in 1927, now carries the traffic between Norwich and Fakenham.

River Wensum

The Patent Rolls for October 1405 record a grant of pontage for one year to 'Thomas Heremyte of Attelbryg for the repair of the great bridge and the commonway leading from it to the little bridge.' An account of the 'Old Bridge at Attlebridge' is given by Mr. T. D. Atkinson in vol. xviii. of *Norfolk Archæology*, from which it appears that in 1912 the bridge had one segmental and two pointed arches, built in two orders, both being chamfered, and a stone on the bridge recorded the rebuilding in the year 1668. The present bridge is built of concrete.

On Corbridge's map of Norwich, published in 1727, five bridges are shown across the river Wensum, called respectively Coslany, Black Friers, Fy, White Friers, and Bishops Bridges.

Bridges in Norwich

Coslany Bridge crossed two branches of the river, and evidently both of these existed in the thirteenth century, as the Records of the City of Norwich refer to the lease of 'a messuage, in the parish of St. Lawrence, which lies between the two bridges of

Coselanye.' Blomefield stated that Coslany Bridge was possibly built of freestone at the end of the sixteenth century, but there does not seem to be any confirmatory evidence. It is now called St. Miles Bridge.

The present Black Friars or St. George's Bridge was built in Portland stone by Sir John Soane in the year 1784, and the details of its construction, given in *Excursions through Norfolk,* although rather technical, are of interest. Amongst them it is stated that 'The voussoirs of the arch have their joints worked perfectly smooth, and are set in dry milled lead. In the middle of each voussoir, two tubes of iron, of three pounds weight, are inserted, and let equally into each stone; the channels are sunk from the tails of the voussoirs to the cavities of the iron joggles, and both of these run full with lead; the whole is ornamented with iron railings.' The full specification is preserved in Norwich Museum.

This bridge took the place of one having three arches, built near the end of the sixteenth century, and towards which Alderman Edward Wood gave the sum of 20 marks. Before this date there had only been timber bridges, one built in the time of Henry V and another about fifty years later. It was then known as 'New Bridge.'

The City Records also include an agreement concerning Fye Bridge, dated August 1st, 1283, in which the Community of the Citizens of Norwich appointed Walter de Monton to be 'the collector of various rents belonging to Fibridge, and all legacies and chance sums left for the sustentation of the bridge on condition that he maintained it so that all passengers might have free and convenient passage.'

The first stone bridge on this site was built in the reign of Henry IV, and had two arches. It was broken by flood in February 1570 and rebuilt three years later. According to Blomefield, it had 'a large and small Arch, the large one 26 feet wide.' Over this arch was an inscription: '1572, Robart Sucklyng Mayor. 1573, Mr. Thomas Peck Mayor. Peter Peterson Chamberline.' The present iron bridge was built in 1829.

Whitefriars Bridge, formerly called 'Seynt Marteyns brigge,' was submerged by the great flood which occurred in the year 1290, and a later bridge was demolished by the Earl of Warwick at the time of Ket's rebellion, in the middle of the sixteenth century, and its timbers were used to strengthen Pockthorpe and Bishop's Gates. In 1591 it was rebuilt in freestone, but the present bridge is comparatively modern.

The only mediæval bridge now remaining in Norwich across the Wensum is Bishop Bridge (Fig. 65), which has three segmental arches having a total span of 24 yards. Originally each arch had five ribs, but only parts of them now remain, except in the case of the middle arch, which still has three complete ribs. The arch-rings are built of stone and brick, but the main structure consists of flint and stone rubble. The width between the parapets is 15 feet, and the semicircular recesses, corbelled out above the cut-waters nearest the city, are said to mark the site of the two angle turrets of the 'Bishop's Gate' that formerly stood at this end of the bridge.

It is recorded that in the year 1275 a Patent was granted to the Prior ' to make what gates he pleased out of the Monastery and to open and shut them and keep them locked at his pleasure; and also to erect a gate with a Bridge 20 foot broad thereto adjoining.' This grant is taken by Blomefield to refer to Bishop Bridge.

Another document quoted in the *Records of the City of Norwich* is a charter to Richard Spynk in the year 1343 recording the various works undertaken by him. This included the building of the gate on the Bishop Bridge, ' et tout les arches de piler enpiler oue le pount tretoz,' a phrase which it is difficult to interpret.

An engraving of Norwich in the year 1741 by Samuel Buck shows a large square tower extending over the entire arch nearest the Cathedral. Corbridge, however, in 1727 only depicted a gateway between two small towers, but his drawing, being on a map, may be only diagrammatic. Wallis's engraving, published in 1818, shows the bridge without any towers much as it stands today.

Three new bridges were constructed in Norwich early in the nineteenth century. A stone bridge called Carrow Bridge was built in the year 1810, and Duke's Palace Bridge, of cast iron, in 1822. Foundry Bridge, of timber, was opened in 1814 as a toll bridge, and remained until 1844, when it was replaced by one of iron.

In addition to the above-mentioned bridges, all of which are across the river Wensum, there is a stone bridge over a channel near the Horse Fair said to date from the end of the thirteenth century. One end is blocked up, but the bridge carried St. Faith's Lane, and, at one time, the precinct wall of the Grey Friars. An illustration of this bridge is given in vol. x. of *Norfolk Archæology,* published by the Norfolk and Norwich Archæological Society.

Below Norwich the only means of crossing the Yare is by ferry or railway until Great Yarmouth is reached. A few miles above Great Yarmouth Bridge, which is quite modern, the Yare is joined by the river Waveney.

The Waveney throughout its entire course forms the boundary between Norfolk and Suffolk. As mentioned earlier in this chapter, it rises close to the source of the Little Ouse, being separated, according to Blomefield, by 'no greater Division than 9 Feet of ground.' It has unfortunately no interesting bridges, the majority being built of iron and the remainder of brick. Scole Bridge bears the date 1838, and Syleham Bridge that of 1847.

River Waveney

The rights of fishing 'from Bungay Bridge to the Earl's Vineyard' were granted in the year 1240 by the Earl of Norfolk to William de Pirnhoe, and in 1352 it was presented that the bridge between Bungay and Ditchingham and also the one from Bungay to Earsham had been 'overturned and submerged by heavy rains.' It was arranged that the lord of the manor and castle of Bungay must supply the timber, and that the men of the town should carry it from his wood and do the repairs at their own cost. The 'bylding of the chapel of our Lady on the brygge in Bungay' is recorded in the year 1532, but the chapel was demolished in the year 1733. According to the Rev. Alfred Suckling, this chapel was situated at the east end of the bridge, on the southern side of the river. The Parish Book of Earsham has an entry for November 11, 1737, ' paid to William Colings for building the bridge between Bungay and Earsham £17, but the town had £5 from the county—£12.' It was called Cock Bridge after the public-house which formerly stood near this place.

A bridge evidently existed at Beccles in the thirteenth century, as mention of one was made in a grant of ' free fishing ' obtained by the Abbot of Bury in the year 1268. There is also a record of the rebuilding of a bridge in the middle of the fifteenth century, several legacies having been left for this purpose. In one case money was left to the Chapel on the ' Great Bridge of Beccles,' which was dedicated to ' the Blessed Virgin Mary.'

The last bridge across the river Waveney is called St. Olave's Bridge, after the Priory of that name. There is a record of a ferry here in the year 1358, although an inquisition taken in 1296 to enquire ' what detriment it would be to anyone if leave were granted to Jeffery Pollerin of Yarmouth to build a bridge at St. Olave's Priory ' gave the verdict that ' it would be detrimental to the holder of the ferry and to the prior of Toft, but it would be to the great benefit of the country.' No bridge, however, was built until the beginning of the sixteenth century, but the Patent Rolls record that leave ' to build a bridge ' was granted in the year 1420. According to Suckling, there was an inscription in Loddon Church, about seven miles west of St. Olave's, which stated that the causeway and bridge were constructed

63. CRINGLEFORD BRIDGE.

64. NEWTON FLOTMAN BRIDGE.

65. BISHOP BRIDGE, NORWICH.

at the sole expense of Dame Margaret, the wife of Sir James Hobart, who died in the year 1522. This bridge, which had three arches, was rebuilt in the year 1768, and Suckling, in 1846, mentioned that the bridge that then existed was 'steep and narrow.' He also stated that before the reign of Edward I the ferry was kept by Sireck, a fisherman, who received 'bread, herrings and such like things' to the value of 20s. a year.

THE HUNDRED RIVER AND THE RIVER BLYTH

These two streams, together with the Minsmere River, drain the northern part of East Suffolk, and although they are crossed by numerous bridges, none are of any archæological interest. The oldest one appears to be Walpole Bridge across the Blyth, adjoining Walnut Tree Farm, two miles south of Halesworth, which has the date 1778 carved on the keystone of the upstream side. It has a single brick arch of about 7 yards span, and the width between the parapets is 12 feet. At Blyford, a few miles below, is an attractive bridge (Fig. 66), also having a single brick arch. In this case the width is but 10 feet.

THE RIVER ALDE AND THE RIVER DEBEN

The south-eastern part of Suffolk is drained by the river Alde, which rises near Badingham a few miles north of Framlingham, and by the river Deben, whose source is only about ten miles farther southwest. None of the bridges over the Alde are ancient, but one over the Deben between Farnham and Stratford St. Andrew, called Stratford Bridge, has the year 1804 carved on the keystone, and Snape Bridge, a few miles below, bears the date 1802. Both bridges have single brick arches with stone keystones. Three bridges of this type still remain across the river Deben, but practically all the others are quite modern.

THE RIVER GIPPING

The river Gipping, which below Ipswich is called the river Orwell, rises a short distance north-west of Stowmarket. Except for a few iron bridges in or near Ipswich practically all the bridges are built of brick. None are ancient, but several notes of interest about Stoke Bridge in Ipswich will be found in the *Annalls of Ipswche*, written by Nathaniel Bacon, Town Clerk of Ipswich in the middle of the seventeenth century, and published in the year 1884. As was often the case, its use was only permitted when the ford was impassable, and a regulation to this effect, made in the year 1495, ordered that 'All carts going over Stoke Bridge, lately built, shall pay towards the repayring and maintaining of the same; viz. every Burgess 1d. o.b. (one penny half-penny) and every forrainer 1d. for ever, provided that none shall goe over the bridge when they may goe through the ffoord.' The bridge was evidently rebuilt in the year 1559, as there was an item for the 'carriage of 28 lodes of timber from Writton to Ipswich for the building of Stoke Bridge.' Even then, it was presumably only a wooden bridge.

Bourn Bridge, over the Belstead Brook, which carries the road from Ipswich to Manningtree, has one stone arch with a projecting keystone on the upstream face. It was widened on the other side in brick and stone in the year 1891, and now has a roadway about 30 feet in width. Bourn Bridge was mentioned in a perambulation made in the year 1352.

THE RIVER STOUR

Rising just within the borders of Cambridgeshire, about seven miles south of Newmarket, the river Stour flows in a south-easterly direction as far as Wixoe, where it is joined by a branch from the south-west coming from Steeple Bumpstead.

Below Wixoe the Stour, for the greater part of its course, divides the counties of Suffolk and Essex, flowing through Clare, Sudbury, Bures and Nayland, to Higham, where it is joined by the river Brett. Above Higham none of the present bridges across the Stour, or its tributaries the Glem and the Box, are of interest, being mostly of brick, built during the nineteenth century. There are also a number of iron or concrete bridges.

Pontage was granted in May 1331 for the repair of Sudbury Bridge, and it is recorded on November 4th, 1520, and again in 1594, that the bridge between Sudbury and Ballingdon had been swept away by flood. The sum of £100 was spent by the County of Suffolk, on their part of the bridge, in the year 1598.

The river Brett, however, which rises a few miles *River Brett* north of Lavenham, has, in addition to a number of quite ordinary brick or iron bridges, several of more than usual interest. At Chelsworth the river is crossed by a brick bridge, part of which was built in 1754, and a few miles below, at Ash Street, there is a bridge (Fig. 67) having three pointed arches with stone vaults, each with three wide brick ribs. The arch-rings, which are chamfered, are built of narrow bricks, as are the faces and parapets of the bridge. The total span is 15 yards, and the roadway is slightly less than 11 feet wide. There are massive cut-waters on both sides of the bridge.

The bridge at the northern end of Hadleigh, which carries the road to Stowmarket, has six iron arches on brick piers. It was built, according to Mr. H. R. Barker, in the year 1843 at a cost of £1,150.

Toppesfield Bridge   Toppesfield Bridge (Fig. 68), at the southern end of Hadleigh, is certainly the finest bridge in Suffolk, having three pointed arches, each with six chamfered ribs. These, and the inner arch-rings, are built of stone, but the outer rings and the faces and parapets of the bridge are of brick. The total span is about 16 yards, and the width between the parapets is 21 feet, the bridge having been widened in brick by about 10 feet on the downstream side. Unfortunately nothing appears to be known concerning the history of this delightful bridge, which probably dates from the fourteenth century. In the *History of Hadleigh*, written by the Rev. Hugh Pigot, published in 1860, mention is made, in the reign of Edward I, of lands called 'Priestisbriggelond and Briggeslond,' which might possibly refer to this bridge.

Cattawade Bridge   About five miles below its junction with the Brett the Stour reaches the sea at Seafield Bay. Close to its mouth the road from Manningtree to Ipswich crosses by a bridge, known as Cattawade Bridge (Fig. 69), which has three semicircular arches span-

66. BLYFORD BRIDGE.

67. ASH STREET BRIDGE.

68. Toppesfield Bridge, Hadleigh.

ning a distance of 24 yards. It is built of brick, but has stone arch-rings, and a stone string-course at the road level. The width between parapets is 15 feet. Walter de Suffield, Bishop of Norwich, who died in the year 1256, left the sum of 2 marks towards the repair of 'Cattawade Bridge,' and in 1350 Roger de Kenton, warden of the bridge, applied for a grant of land 'on the high road, to himself and his successors,' in order that he might build a chapel by the bridge. The confirmation of this grant is recorded in the Patent Rolls of June that year, and it was specified therein that the plot of land was to be ' 100 feet long and 48 feet broad, parcel of the King's highway by the causey of the bridge, and of a value of one half-pence yearly and no more.' The chapel was to be dedicated to the Virgin Mary. Ten years later protection was granted, to the hermit of ' Cattiwade,' to seek alms towards the maintenance of the bridge and causeway, and the Quarter Sessions Records for the County of Essex include many entries of money spent on the repair of Cattawade Bridge, including £105 in the year 1653 and £250 in 1680.

THE RIVER COLNE

Although more than twenty bridges are shown over the river Colne on J. Chapman's map of 1774, it is doubtful if any of the existing bridges, except the one that crosses about half a mile south of Castle Hedingham, are more than a hundred years old; in fact, more than half of them are now girder or concrete bridges.

Castle Hedingham Bridge has three brick arches with stone keystones, spanning a distance of 12 yards.

From an inscription on the bridge it appears that it was built in 1736 and 'enlarged' in 1819. The width between its parapets is now about 28 feet, and according to the *History of Essex*, written by Peter Muilman in 1770, this bridge was built by Sir Henry Houghton, Bart., in place of a wooden one.

New Bridge (now a girder bridge) on the road from West Bergholt to Lexden, about two miles west of Colchester, was, according to Mr. J. H. Round, known as 'novus pons' as early as the year 1204.

Bridges in Colchester

Three bridges cross the river Colne in Colchester. Two of these, the North and East Bridges, were described by Morant, in 1748, as being built of timber, while Hythe Bridge was then of brick. North Bridge was rebuilt, with three brick arches, early in the nineteenth century by Sir William Staines, Lord Mayor of London, but has since been replaced by an iron bridge. The East Bridge, rebuilt about the year 1816 with five elliptical arches having projecting keystones, was widened on each side in concrete about five years ago. The present Hythe Bridge, which dates from the year 1878, took the place of one built in 1737, but there were, however, several earlier bridges on or near this site; the earliest on record was a foot-bridge, built about the year 1407. According to an 'Oath Book,' quoted by Morant, the bailiffs and Council of the Town covenanted that this bridge 'should not be above eighteen inches wide; and never be made fit for horses or carts to go over. If it were otherwise, or proved any prejudice to this town, it was immediately to be demolished, and it was to be so built as not to hinder the Navigation up to East bridge.'

About seventy years later the need for a wider bridge was evidently imperative, for the Curia Regia Rolls for the year 1474 mention an indenture by which it was arranged ' to make a brigge of Stone or Tymbyr, or of bothe, over the said Haven, Rever, and Water for men, hors, and carte to passe there over to and fro for ever, with a Draughte (drawbridge) in the same, that Shippez, boytez and oder Water-vessellez shall mowe passe there, if the Water will serve therfore.'

About three miles below Colchester, the Colne is joined by the Roman River, which rises two miles to the north-east of Coggeshall. Most of its bridges have been rebuilt within the last forty years, but Stanway Bridge, which carries the road from Colchester to Coggeshall, may date from early in the nineteenth century. Fingringhoe Bridge, like many other timber bridges, was a continual source of trouble, as will be seen from the paper by the Rev. G. Montagu Benton, F.S.A., in vol. xx. of the *Transactions of the Essex Archæological Society*, which gives many interesting extracts from the Quarter Sessions Rolls. It has, unfortunately, been recently rebuilt in concrete. *Roman River*

THE RIVER BLACKWATER

The river Pant, as the Blackwater is called above Becking, rises near Saffron Walden in the extreme north-west of Essex. It is crossed by numerous bridges, but it is doubtful if any of the existing ones were built before the beginning of the nineteenth century except the one, known as Long Bridge, on the road from Coggeshall to Kelvedon. This bridge *Long Bridge*

(Fig. 70), which has three obtusely pointed brick arches, is probably mediæval. It was widened downstream in 1912 in brick and concrete, and the parapet on the other side is now carried on wooden beams resting on the tops of the massive cut-waters, giving a total width between the parapets of about 23 feet. The total span is 13 yards.

Blackwater Bridge   A bridge between Coggeshall and Braintree, described as being at Stratford and on or near the site of the present Blackwater Bridge, where Stane Street crosses the Blackwater, was the cause of much recrimination in the fourteenth century. The Patent Rolls for October 1341 record a decision that 'the abbot of Coggeshall be quit of the repair and the maintenance of the bridge,' based on evidence, given in the reign of Edward II, that 'from time of memory there has been no bridge other than a wooden plank on which passers by have been able to cross safely.' Many years earlier, in 1284, an inquisition had decided that 'Nobody of right is bound to repair the bridge of Stratford, between the market of Coggeshall and Braintree, because, when it has been damaged it has been repaired by alms'; but another inquisition, taken in June 1308, reversed this decision: hence the appeal made in 1341.

In the Muniment Room of Westminster Abbey is an interesting document and plan (Nos. 6578-9) concerning a claim by the Crown that the Abbot of Westminster, the Abbot of Coggeshall, and others holding fishing rights in the river, should rebuild the wooden bridge, shown on the plan as 'Blakwaterbrigge,' which in the year 1520 had been 'totaliter diruptus confractus et prostratus' by flood. It is interesting to find that this bridge bore its present name as early as the sixteenth century.

At Maldon the Blackwater is crossed by two bridges. The one over the northern channel was called the 'High Bridge' by Morant, in 1768, and was described by him as a bridge 'consisting of five arches and which seems to be very old.' To the second one, 'over what is now the main stream,' he gave the name of Fulbridge. Ralph Breder, Alderman of Maldon, who died in March 1608, left the sum of £120 towards 'repairing the Haven, Channels and the Bridges of Fullbridge and Heybridge.' The Patent Rolls contain an interesting and very unusual entry concerning the latter bridge, for, in November 1407, in aid of 'the great expense sustained by them in the repairing of a bridge called Hebregge, by the town, destroyed by innundation of the sea,' the burgesses of the town of Maldon, co. Essex, were allowed to 'be quit of coming to any Parliaments for seven years, so that they faithfully apply the costs of coming to the repair of the bridge.'

Bridges at Maldon

THE RIVER CHELMER

The Chelmer, which is now joined to the river Blackwater by a canal, also rises near Saffron Walden. It runs through Thaxted and Dunmow to Chelmsford, where it is joined by the river Can. Both bridges at Great Dunmow are mentioned in the Essex Bridge Books for the sixteenth century; the one called 'Church End Bridge' was reported in 1562 to have 'fallen,' and a few years earlier 'Park Bridge' was said to be in decay. The former was described as being a timber bridge, and it is probable that Park Bridge was also of this type.

The bridge over the Chelmer between Chelmsford and Springfield, which carries the road to Colchester, and shown as Springfield Bridge by Thomas Yeoman in 1765, was reported to the Sessions of 1562 as being 'in such a state of ruin through defects of the timber as to be dangerous to foot and horse passengers.' The responsibility for its repair devolved on the Queen and Sir Henry Tyrryll.

Moulsham Bridge

The present bridge across the river Can, which connects Chelmsford and Moulsham, was built in the year 1787 to the design of Mr. Johnson. It has one segmental stone arch, with a span of 14 yards, and stone balustrades, which are about 30 feet apart. As will be seen from the illustration (Fig. 71), it is an unusually attractive bridge.

The first bridge between Chelmsford and Moulsham is said to have been built by Maurice, Bishop of London, in the reign of Henry I, and in 1351, when the bridge was said to be broken, it was stated that Ralph, Bishop of London, and the Abbot of Westminster were bound to do the repairs. The Treasurer's Roll (W.A.M. No. 19866), which is preserved among the Muniments of Westminster Abbey, records the expenditure, in the year 1372, of £23 6s. 8d. on building a new bridge at 'M'lsham' by Henry de Yeuele, the master builder who built the present nave of the Abbey. Another document (W.A.M. 31840) gives details of the cost of repairs undertaken in the year 1520, from which it appears that three carpenters were employed in addition to the master mason and a man whose work was the 'laisshyng and hewyng of tymbre.' One of the earliest items in these accounts was: 'For the Master Mason of the Kynges Workes labor to se the bridge and to have his counsell and his brekfast—xijd.' Mention was made of wrought iron 'for the gyne' (crane) and for 'cartyng of tymbre,' but no reference was made of any stone or lime, so that the bridge was evidently built entirely of timber.

Although the County of Essex has now few road-bridges of archæological interest, several mediæval bridges still remain leading to moated houses. The one at Pleshey Castle, which has a single pointed brick arch, is said to date from the fifteenth century, and another (Fig. 72) leading to Latchleys Manor House, of slightly later date, has two arches spanning a distance of about 8 yards. The width between its parapets is 9 feet, and this bridge is also built entirely of very narrow bricks. The bridge at Tolleshunt D'Arcy Hall, which has four semicircular arches, is built of brick and stone and bears the date 1585, and remains of a bridge can still be seen in the moat at Halstead Rural.

The rivers Roding and Lea, being tributaries of the river Thames, will be described in the following chapter.

CHAPTER VIII

THE NORTHERN TRIBUTARIES OF THE THAMES

THE RIVER RODING

THE river Roding has its source at Chapel End about four miles north-west of Great Dunmow Above Chipping Ongar, where it is joined by the Chipsey Brook, all the bridges are comparatively modern, and the earlier bridges on these sites were in most cases built of timber.

Passingford Bridge Five miles below Chipping Ongar the road to London crosses the Roding by a brick bridge called Passingford Bridge, built in the year 1785. It was reported to the Sessions of 1567 that 'Pyssingfordes Bridge conteyninge in lengthe by estimacion xxxiiij footes which is in a common way to London' was in great decay.

The bridge at Abridge and also Loughton Bridge, about two miles below, are both built of brick with stone keystones. They probably date from early in the nineteenth century.

Woodford Bridge Woodford Bridge (Fig. 73), which bears the date 1771, is of more elaborate design, having stone voussoirs, string-courses, and pilasters on the outer faces of the parapets. The total span of the three arches is 16 yards, and the width between the parapets is nearly 20 feet. Numerous entries concerning Woodford Bridge will be found in the Essex Bridge Book and the Rolls of the Quarter Sessions, and from these it appears that in the year 1566 the bridge was 'ruinous.' In the second year of the reign of James I it was recorded that 'it was but a horse bridge, and the country were desirous to have it made a cart bridge.' It is difficult to say when the first brick bridge was built on this site, as in the year 1674 a bill for £19 4s. 6d. was presented which included 'bricklayers work,' while twenty-one years later Andrew Pitts, carpenter, was paid the sum of £130 for repairing Woodford Bridge. It may have been a timber bridge with brick abutments.

THE RIVER LEA

The river Lea rises in Bedfordshire, about three miles north-west of Luton. According to Mr. W. Austin, in his *History of Luton*, there were eleven bridges in that town in the eighteenth century, but they were evidently quite small ones, as even in the year 1795 there was only a footbridge on the Bedford Road.

Below Luton the Lea is still fordable in many places, and none of its bridges above Hertford are ancient, although in many places the river has been bridged since early times. The bridge of Stanborough and Stoken Bridge were mentioned in a 'grant of free fishing' made in the year 1277. At Hatfield, however, there was only a ferry in the reign of Edward II, the tolls of which, together with those of the bridges of Ware and Thele, belonged to Aymer de Valence, Earl of Pembroke.

In Hertford, the Lea is joined by the river Maran and the river Beane. The former comes from Welwyn and the Beane from Walkern. None of the

Rivers Maran and Beane

138 ANCIENT BRIDGES

bridges over the Maran are of interest, but at Watton at Stone the road to Walkern crosses the river Beane by a stone bridge (Fig. 74) which, although of no great age, is notable, as it is the only stone bridge in the district. Its three segmental arches span a distance of 10 yards, and the roadway is 15 feet wide. According to the Records of the Quarter Sessions, there was only a foot-bridge on this road in the year 1673, and this was then in a 'decayed' state and 'very dangerous to strangers.' In 1838 it was recorded that both Watton Moor and Watton New Bridges were built of brick.

Bridges in Hertford

In Hertford, the bridge across the Beane is known as Cow Bridge. It was of wood until the year 1676, when it was rebuilt in brick. According to Mr. Lewis Turnor, who wrote his *History of Hertford* in 1830, it had then been 'lately repaired and much improved.' The map in his *History* shows also the Castle and Mill Bridges over the river Lea, and he recorded the fact that Folly Bridge, leading from the Bull Plain to the Folly, was first built in the year 1738.

River Rib

About a mile below Hertford the river Rib joins the Lea. It rises a few miles north-west of Buntingford, where it is crossed by a brick bridge having stone keystones, which bears the date 1766. The sum of £1 6s. 8d. was left in 1494 by Ellen Barbour for the repair of the bridge 'in the Chapel end' in Buntingford, and in May 1682 the repairs to 'Chappell Bridge' cost the sum of £2 16s. 9d. Two years later a further £5 17s. 0d. was spent. It appears that the bridge was then built of timber, and even in 1796 the bridge had a 'wood coping' which was said to be out of repair, although the arches were stated to be 'perfectly sound.'

Ford Bridge at Braughing, which carries the road from Ware to Cambridge, also has two brick arches. At the end of the sixteenth century Ford Bridge was said to be 'ruinous,' and in the year 1659, when described as 'a cart-bridge,' it was 'in great decay.' At Standen, the next crossing, there is now an iron bridge, but an interesting note is to be found in the Hertfordshire Sessions Rolls of the year 1590 to the effect that 'We fynd that ther is a brydge decayed in the parishe of Standen, called Our Lady Bridge, which is in the higheway to Starteford and is in great ruene and deceye, which had some time a Chapple upon yt.' Throughout the eighteenth century it was often reported to be in the same state of disrepair, and in 1782 it was suggested that the bridge should be 'rebuilt with brick sufficiently wide for carriages to pass over.' It was then described as 'a wooden bridge,' and estimates were obtained the following year for building a brick bridge with five arches.

Even at the present time there are several fords across the river Rib below Standen, which at times are very deep and dangerous, but at Wadesmill, where Ermine Street crosses the river, there is a bridge (Fig. 75) of very unusual design having a row of pillars, six in number, instead of a central pier. There are only two other examples of this construction known to the author. One is over the river Creedy in Devon, and the other crosses the Canal on the northern side of Regent's Park, close to the Zoological Gardens. The present bridge at Wadesmill was evidently built about the year 1825,

Wadesmill Bridge

as a letter, dated at Wades Mill on October 15th in that year, stated that the new bridge was 'now arched over,' and in 1838 the bridge was described as of 'brick and iron.' In the seventeenth century it was only a wooden bridge, and its rebuilding in brick was recorded in 1675. Four years later, however, the middle pier sank, having been undermined by the stream, and had to be partly rebuilt. In 1767 the bridge was reported as being 'out of repair and too narrow for the passage of waggons.' At that time it had five arches, and a proposal was made that two of them should be closed up.

It is interesting to note that the turnpike at Wadesmill, erected in the year 1662, was one of the first three in England, the others, according to Cussans, being at Caxton in Cambridgeshire and at Stilton in Huntingdonshire. Ten years later the receiver of the tolls at 'le Turnepike' at Wadesmill was indicted for 'converting to his own use divers sums of money received by him in virtue of his said office.'

The river Lea was made navigable up to the town of Hertford under an Act passed early in the reign of George III. Close alongside is the Lea Navigation Canal and another channel, called the New River, which was made early in the seventeenth century to provide a water-supply for London. Timber bridges were shown by Ogilby across both rivers at Ware, and it is probable that all earlier ones were of this type. In the year 1191 the bridge of Ware was broken by the men of Hertford in order that all traffic should cross the Lea in their town. The bailiff of Hertford claimed rights both for the use of the bridge and the adjoining ford. Nearly seventy years later the bridge was again broken and a channel was dug through the ford to prevent its use. After a few years, however, the bailiffs of Ware succeeded in 'turning aside the high road which used to go from Hertford to Ware to the detriment of the vil of Hertford.'

About two and a half miles below Ware the road from Stanstead St. Margaret's to Stanstead Abbots now crosses by iron and concrete bridges. According to the *Victoria County History*, this crossing, in the twelfth century, was called 'Pontis de Thele,' and in the following century the dues from this bridge were taken by the warden of Hertford Castle. It was a wooden bridge until the year 1873.

River Stort

The river Stort, which joins the Lea two miles below Stanstead, rises about six miles south-west of Saffron Walden. It runs through Bishop's Stortford, Sawbridgeworth, Harlow, and Royden, and was made navigable in the middle of the eighteenth century. None of its bridges are earlier than that period, and most of them have been rebuilt during the last hundred years.

Bridges at Waltham

Leland recorded that 'there be 7 or 8 Bridges in the Toune of Waltham. For there be divers Socours of Streamelettes breking out of the thre principalle Partes of Luye Ryver,' and at the present time there are still numerous bridges. None, however, is ancient except the one situated about 300 yards N.N.E. of the Abbey Church, which is in a very derelict condition. Fortunately it is now fenced in and safe from wanton damage. The single arch, three-centred in shape, still has three ribs, and it is probable that at one time there were six, but the other three have been destroyed.

According to an article written in March 1859 by the Rev. Edmund Littler, and printed in vol. ii. of the *Transactions of the Essex Archæological Society*, 'the ribs are broad and chamfered, the joggles cemented with lead . . . and the span is about 18 feet.'

The Patent Rolls for April 1310 record a grant of pontage for a period of three years for the repair of the bridges between Eustacebrigge and Waltham Cross, and a further grant was made in February 1336.

Bow Bridge The famous bridge across the Lea at Stratford was built by Queen Matilda, wife of Henry I, who, according to Stow, 'caused two stone bridges to be builded in a place one mile distant from the Old foord, of the which, one is situated over Lue at the head of the towne of Stratford, now called Bow, a rare piece of work, for before the time the like had never beene seene in England.' This fact was evidently obtained by Stow from an inquisition taken in the year 1248 which is quoted in *Archæologia*, vol. xxii., and it appears that the Queen purchased 'certain lands, rents, meadows and one water mill, called Wiggen Mill,' for the support of the bridges and causeway, which she gave into the charge of the 'Abbess of Berking.' At a later period, however, after the Abbey of Stratford had been founded, the lands, etc., were leased to this abbey, which undertook the responsibility for maintenance of the bridge.

The bridge, destroyed in 1836, had three stone arches which, according to Plate XI. in *Archæologia*, vol. xxvii., were four-centred in shape, widened with segmental arches. The width between the parapets had originally been 13½ feet, but in 1741 it was increased to 21 feet. The middle arch had no ribs, but each of the others had a single rib, 4½ feet in width, a very unusual feature. In the reigns of Henry VI and Edward IV this bridge appears to have had a chapel dedicated to St. Katharine. It was probably rebuilt about the fifteenth century, as the four-centred arch was not in use as early as the reign of Henry I.

THE RIVER FLEET

This stream, whose existence is unknown to most people, as it is now covered over, was, in mediæval days, an open river crossed by several bridges, some of which are mentioned in ancient documents.

The Rolls of Parliament for the year 1306 record a petition, presented by the Earl of Lincoln, which stated that 'the water-course under Holbourn and Fleet bridges used to be wide and deep enough to carry 10 or 12 ships up to Fleet bridge . . . and some of them passed under that bridge to Holbourn bridge.' It was petitioned that 'the lord mayor, with his sheriffs and discreet aldermen, may view the watercourse . . . and cause all nuisances thereon to be removed.'

In November 1307 a commission was appointed 'to survey the river Fleet from Holeburne Bridge to the Thames; its course being so obstructed by filth cast into the stream as well as by a quay . . . built upon the bank of the Thames by Baygnard Castle.'

In the time of Stow, Fleet Bridge was 'of stone, coped on each side with Iron Pikes.'

THE TYBOURNE AND THE WESTBOURNE

The upper part of the Tybourne, or Aybrook, formed an important water-supply for the City of London as early as the year 1236, when lead pipes were laid to convey its water from Tyburn to the City.

According to Mr. Charles T. Gatty, in *Mary Davies and the Manor of Ebury*, the Manor of Eia, which from the Conquest until the reign of Henry VIII was held by the Abbey of Westminster, was bounded on the east by the Tybourne and on the west by the Westbourne. The Tybourne flowed under Oxford Street from the west side of Stratford Place into Davies Street, then down South Molton Lane and Avery Row. In Brook Mews, below Claridge's Hotel, a 'pier wall was laid bare with iron rings for mooring boats.' It was crossed at the Piccadilly end, a little east of Brick Street, by the Stone Bridge which gave the name to a close, or field, shown on the plan of 1585, which forms the basis of Mr. C. L. Kingsford's book entitled *Early History of Piccadilly, Leicester Square and Soho*. Below this bridge it passed through the site of Buckingham Palace, there dividing, one stream going to Westminster, where it further divided and formed the Isle of Thorney, on which the Abbey was built, and the main river, passing on the west side of the Stag Brewery, flowed down Tachbrook Street. Abbot's Bridge at the intersection of Warwick Street and Tachbrook Street was shown on a map of 1662, but not on the plan which shows the manor as it was in 1614.

69. CATTAWADE BRIDGE.

70. LONG BRIDGE, COGGESHALL.

71. MOULSHAM BRIDGE, CHELMSFORD.

72. LATCHLEYS MANOR HOUSE BRIDGE.

THAMES—NORTHERN TRIBUTARIES

The Westbourne was shown by Ogilby as passing under the road to Oxford, now called the Bayswater Road, by a 'Brick Bridge' at the 'Watering Place' and the road to Hammersmith at 'Knight's Bridge.' The latter bridge appears to have been close to the present Albert Gate, and the stream now flows underground along the east side of Lowndes Square, crossing over Sloane Square Station in a large cast-iron pipe and joining the Thames near the Chelsea Suspension Bridge.

There had evidently been a bridge at Knightsbridge since early in the Middle Ages, as in the year 1380 pontage was granted to Jo. Crowcher, of Knyghtebrigg, towards repairing and amending the King's highway from London to Braynford, the tolls to be taken 'at Knyghtebrigge or elsewhere.' In 1825 it consisted of two brick arches under the road leading into an arched conduit 80 feet in length.

The Churchwarden's Accounts of Melton Mowbray contain an item for the year 1598 which reads: 'Payd to an aspitall cauled Knightbridge in Middellsekes the 8 of Meaye—vi d.'

About the year 1399 it was presented that 'there is a wall by the Thames at Westminster, below the place of William Henney, where time out of mind there was an arch four feet wide and more for the ebb and flow of the Thames into a ditch called Quenesdyche . . . and that the said arch was stopped with stones and mortar by Nicholas Slake, dean of the college of the King's chapel of Westminster, so that the water of the said river could not ebb and flow into the said ditch.'

An earlier presentment by a jury of Ossulstone

hundred stated that 'a little bridge at Charing Cross is so stopped with filth, namely dirt and sand, that the water cannot run through it in time of rain.' This bridge also appears to have been over the Quenesdyche.

THE RIVER BRENT

The river Brent rises near Hendon, its many tributaries forming the Brent Reservoir at the Old Welsh Harp. One tributary is crossed by Watling Street at Edgeware, and it is interesting to find the record in the Coram Rege Rolls of a presentment, made in the year 1370, that 'the prior of St. Bartholomew, London, ought to repair a wooden bridge called Eggewerebrigge and that his default has made the King's road impassable for a year past.' The Prior resisted the claim; the jury gave the verdict in his favour, stating that this bridge 'was from all time repaired by alms of the men of the country and others crossing it.'

During the fourteenth century many bridges over the river Brent were evidently in a bad state, judging from the number of complaints recorded in the Coram Rege Rolls. Many of these cases are given very fully by Mr. C. T. Flower in his books on *Public Works in Mediæval Law*, published by the Selden Society.

In the year 1339 it was presented that 'the abbot and convent of Westminster ought to repair Stickledon bridge over the Brent between Hanwell and Greenford for foot passengers and horse men.' It was stated that the bridge was first built in the reign of Henry III to give access to the windmill, but had been broken for twenty years, and the former Rector of Perivale and two others had been drowned while attempting to cross. In this instance the jury found that the Abbot was responsible.

In 1398 the bridge at Hanwell, 'for men with horses and carts,' was said to be broken by default of the Prior of Ogborne, as was also 'Brent Bridge,' between Northcote and Acton.

Ogilby showed a brick bridge with six arches over the Brent close to Hanwell Church, on the road from Acton to 'Norcoat' and Uxbridge, and Leland also mentioned a bridge with six arches about a mile from Southall.

The Roman road from London to Staines crosses at Brentford, and a bridge must have existed here since early times. Pontage for the maintenance of the bridge of 'Brayneford' was granted in 1280, and again in the years 1331 and 1369. According to the *Report of the Committee of Magistrates appointed to make enquiry respecting Public Bridges in the County of Middlesex*, published in 1825, a stone bridge with a single arch was built here in 1824 in place of one made of brick and stone erected in 1740. This Report contains particulars of every bridge in the County of Middlesex, and is a very valuable record.

THE RIVER COLNE

The river Colne and its tributaries, the Var, the Gade, the Chess, and the Misbourne, drain the valleys lying to the south-east of the Chiltern Hills. The majority of the present bridges over these streams are built of brick, the remainder being of iron or concrete, and it is very doubtful if any of the bridges are more than 150 years old. There is, however, a pair of attractive bridges over its tributary, the river Gade, at Waterend, near Great Gaddesden, one of which is shown in Fig. 76. They are built in the mediæval style with pointed arches, but only date from early in the nineteenth century, when the present road was made.

The *Victoria County History for Hertfordshire* gives several notes about early bridges, and it appears that at the beginning of the sixteenth century there was a chapel on the causeway of the bridge at London Colney, a few miles south-east of St. Albans. Leland mentioned a wooden bridge at Berkhampstead, and there appear to have been three bridges in the Manor of King's Langley at the end of the thirteenth century, called respectively 'Longebrygg,' 'Sheffordbrygge,' and 'le Mullebrygge.'

Hunton Bridge, over the river Gade, was reported to the Sessions of 1820 as being 'so built as to render the turning over it at the south-east end very sharp and dangerous for carriages passing over it.' Two years earlier a carriage had overturned and a passenger was killed.

Ogilby's route from London to Buckingham passed through the Chalfonts. He showed the Misbourne as an unusually wide stream and added a note: 'A River that Horses and Carriages go through.' He also showed a 'Wood Bridge' over the Colne at Uxbridge, and another of this type at Colnbrook.

An unusual complaint was reported, at the end of the fourteenth century, concerning the bridge at Stanwell, about three miles north of Staines, for a presentment was made in Easter term 1398 that 'a bridge over the Colne at Stanwell has been newly erected by Ralph atte Mille, where no bridge ought to be, and that the neighbours' animals, namely oxen, calves, sheep and pigs pass across it . . . and eat the corn and grass."

He was amerced the sum of 3s. 4d., a poor reward for his enterprise.

THE CHALVEY DITCH

Although this stream is quite small, it possibly recalls many memories to Etonians. According to the *Victoria County History for Buckinghamshire*, Beggars Bridge, also called Spitel Bridge, connecting Eton and Upton, was in the year 1302 stated to be 'broken down and destroyed to the danger of travellers and to the injury of the adjacent country.' It was also reported that Walter le Teb of Eton had built a bridge of wood over the rivulet, 'where there was no previous bridge,' some fifty years earlier— *i.e., c.* 1250—with the aid of voluntary gifts, and had maintained it during his life, but that there was no obligation for anyone to build or maintain it. Further, it was said that the flood of the Thames had so deepened the stream that in the spring no person on foot or on horseback could get across. This bridge was superseded by a stone one mentioned in a College grant of 1443 as 'Spitelbrigge.'

The Hundred Rolls of 1275-6 refer to 'the villata of Eton from Baldewin Bridge to Windsor Bridge.' These are still the boundary points which divide the town from the College on the North and Windsor on the South.

THE RIVER THAME

From a charter granted in the year 1554, which is quoted by Mr. Robert Gibbs in his *History of Aylesbury*, it appears that there were, in the reign of Queen Mary, four bridges at or near Aylesbury over the river Thame and its tributaries. The one on the road to Hartwell and Thame was called Wall Bridge, that leading to Bicester bore the name Stanne or Gallows Bridge, and Holmans Bridge carried the road to Buckingham. Walton Bridge, however, on the eastern side of Aylesbury, is apparently the only one mentioned in mediæval documents. Pontage was granted in June 1384 and in May 1388 for the repair of Waltonbrugge, but a grant issued in October 1398 was for 'Stanbrugge, Omannebrugge, Waltonbrugge and Spetilbrugge,' near the town of 'Aillesbury.' In May 1410 the grant was made to Edmund Seman and the good men of Walton for the repair of the bridge of Walton by Aylesbury, and was renewed in February, 1414. Wall Bridge, rebuilt in concrete in 1927, was shown by Ogilby as a stone bridge with three arches.

Ickford Bridge   A girder bridge now crosses the river between Thame and Long Crendon, where in Leland's days there was a stone bridge with four arches, but at Ickford, about five miles farther downstream, a seventeenth-century bridge still remains. The northern arch is four-centred in shape, while the middle one is three-centred, and both have double arch-rings. The southern arch is segmental. There is one cut-water on each side of the bridge, and on the south side of the parapet over the eastern recess is carved ' 1685. Here ends the County of Oxon,' and on the north side, ' Here beginneth the County of Buckingham 1685.' According to Leland there was a stone bridge 'of 2 Archis' over the northern branch of the river, and a ' Wood Bridge ' across the other one. In 1237 orders were given for the delivery of one oak from the wood of ' Brohull for the repair of Ickford Bridge.'

The road from Oxford to High Wycombe and London crosses the Thame by Wheatley Bridge, which has three semicircular stone arches with projecting keystones. There is also a single arch under each approach, and the one on the eastern side, being pointed in shape, evidently formed part of a mediæval bridge. It has, however, been widened on both sides. Leland described the bridge as having eight arches, but Ogilby marked it as a ' Stone Bridge and 2 Arches called Wheatley Bridge.' *Wheatley Bridge*

About four miles below Wheatley Bridge a causeway 50 yards in length carries the road from Chislehampton to Stadhampton. It has eight segmental arches, each with double arch-rings, and has been widened on the upstream side by iron girders carried on stone piers. The other face still has massive cut-waters, five of which have recesses for foot-passengers. These now give the bridge a very strange appearance, as the intervening parapets have been replaced by iron railings, leaving the masonry of the recesses standing alone about 3 feet above the road level. *Chislehampton Bridge*

' Cheselhampton Brygwey ' was mentioned in a presentment made in the year 1398, when it was reported that ' the King's road there ' was flooded, so that 'men with horses and carts cannot pass thereby.'

Leland, on his journey from Haseley to Dorchester, rode over a bridge before reaching Drayton which he described as having '5 great Pillers of stone apon the which was layed a Timbre Bridge.' This was probably on or near the site of the present girder bridge called Hayward Bridge. According to him the bridge at Dorchester had '5 principale Arches' and a 'Causey' at each end. Ogilby, however, although he showed the bridge, unfortunately gave no details. In December 1381 pontage was granted to the bailiffs of Dorchester for the repair of their bridge, and early in the following century Sir John Drayton bequeathed money for the same purpose.

THE RIVER CHERWELL

Close to its source, about five miles south-west of Daventry, the river Cherwell is crossed in the village of Charwelton by a mediæval packhorse bridge only 3 feet in width between the parapets. Its two pointed arches have a total span of 16 feet, but their appearance is now spoilt by the modern road which runs alongside and masks the lower part of the arches. There is, however, a delightful engraving of this bridge in Baker's *History of Northants*, published early in the nineteenth century.

Trafford Bridge, which carries the 'Welsh Road' across the river Cherwell, has two semicircular stone arches built in a very finished masonry, and may date from the eighteenth century. Cropredy Bridge, about four miles below, bears the date 1691 on the

73. WOODFORD BRIDGE.

74. WATTON AT STONE BRIDGE.

75. WADESMILL BRIDGE.

76. WATEREND BRIDGE.

capstone of the cut-water on the upstream side. The pointed arch is certainly older than the seventeenth century, and the date probably refers to the rebuilding of the second arch after its destruction during the Civil War. The bridge has been widened in blue brick on the downstream side, and is now 17 feet wide between the parapets.

Leland described the bridge at Banbury as having '4 Arches very fayre of Stone,' and this was no doubt the bridge described by Alfred Beesley in his *History of Banbury*, published in 1848. Beesley gives, however, a different number of arches, as he described the bridge as having two arches at the western or canal end, then two more 'usually dry arches,' and farther east three large arches over the main stream. All were pointed in shape and had ribs. He illustrated the western arch, and from the engraving (Plate XVIII.) it appears that this arch had double arch-rings and five chamfered ribs.

Banbury Bridge

Nell Bridge was shown by Ogilby as having six stone arches, but the present one is modern. It is interesting to find that this bridge was called Neile Bridge in the reign of Henry VII.

According to Jeffery's map of 1776, there were no bridges at that date across the river Cherwell between Nell Bridge and the one at Lower Heyford. The latter, called 'Heiwood Bridge' by Leland, has nine stone arches, five of which are pointed, four still having ribs, but in two of these the spaces between the ribs have been filled in to strengthen the bridge. It is now called Heyford Bridge.

The next one, at Enslow, was called by Leland Emley Bridge. It was shown by Ogilby as a 'Wood

Bridg,' so the present one, built with obtusely pointed arches, could not have been built before the last quarter of the seventeenth century. On the downstream side the bridge has been widened with semicircular arches. There are four corbel heads under each of the pointed arches, evidently to carry the centering used when the bridge was being built.

The bridge on the eastward side of Kidlington, which carries the road from Oxford to Bicester, was called Gosford Bridge by Leland, but shown as Gofford Bridge by Ogilby, who noted that it was a stone bridge. The present one, which has projecting keystones, may date from the end of the eighteenth century. The 'Bridge of Goseford, Co. Oxon in the Manor of Hampton,' is mentioned in the Patent Rolls for May 1319.

River Ray

About a mile and a half below Gosford Bridge the Cherwell is joined by the river Ray. At Islip, the Ray is now crossed by a bridge with three arches built of brick and stone. It was a stone bridge in the sixteenth century, but during the Civil War the bridge was destroyed by General Fairfax. Ogilby, thirty years later, described it as having six stone arches.

Magdalen Bridge, Oxford

As mentioned in the *Ancient Bridges of the South of England*, pontage was granted in October 1328 for the repair of bridges over the Thames and the 'Charwell.' Magdalen Bridge, in Oxford, was formerly known as the East Bridge, for in his will, which was proved in the year 1459, Henry Philyp, Alderman of Oxford, left the sum of 6s. 8d. 'to the reparacion of the Est bryg of Oxford,' and the same name is used in a grant of fishing in the Cherwell made in September 1483. It was evidently known by its present name at the end of the sixteenth century, as Hollingshed recorded that 'a great part of Maudlin bridge was carried away by flood' in 1571.

According to an inscription on the downstream side of the bridge, the foundation-stone of the present structure was laid on March 26th, 1773, and was widened by 20 feet in 1882.

THE RIVER EVENLODE

The river Evenlode rises a short distance from Moreton-in-Marsh, and for several miles follows along the boundary between the counties of Oxford and Gloucester. It is crossed by numerous stone bridges, practically all of the same type, having segmental arches and string-courses at the road level. A few have semicircular arches, mostly with projecting keystones, and it is probable that they were built towards the end of the eighteenth century, but unfortunately none of them bear dates.

Many of the bridges over its tributary, the river Glyme, are of this type, and the only outstanding bridge is the one in Woodstock Park, built in the year 1716, across the lake which was formed by damming back the river. This bridge has one lofty semicircular arch between two smaller ones.

River Glyme

The bridge of Blaydon was mentioned in a Charter of 1235.

THE RIVER WINDRUSH

Rising close to the borders of Worcestershire, the river Windrush flows through Temple Guiting,

Naunton, and Lower Harford to Burton-on-the-Water, where it is joined by the river Dikler, which drains the valley lying a mile west of Stow-on-the-Wold. The majority of the bridges are built of stone and some have projecting keystones. Bourton Bridge (Fig. 77), near Bourton-on-the-Water, which carries the Fosse Way, is built in an uncommon style, having a pointed arch between two semicircular ones. All the arches have chamfered arch-rings and the terminals of the parapets are of unusual design. The roadway is 22 feet in width and the three arches span a distance of 9 yards. This bridge was probably built during the eighteenth century.

New Bridge, near Clapton, about $2\frac{1}{2}$ miles farther downstream, is another bridge built in the mediæval fashion with a pointed arch.

In Bourton-on-the-Water the Windrush is crossed by several bridges, the most attractive of which has three flat segmental arches with chamfered voussoirs, and in the village of Lower Slaughter a stone footbridge of unusual design crosses a tributary of the river Dikler.

Burford Bridge

In Burford, the river Windrush is crossed by a massive stone bridge (Fig. 78) having four segmental arches which have ribs. The arches are so low that it is difficult to see under them, but in the case of one arch the ribs appear to be chamfered. The total span is 25 yards and the roadway is 15 feet wide.

According to *The Burford Records*, by Mr. R. H. Gretton, it was 'presented' in the year 1322 that 'the part of Burford Bridge, lying in the parish of Fulbrook, is in a ruinous condition,' and in July the following year pontage, for a period of three years,

77. BOURTON BRIDGE.

78. BURFORD BRIDGE.

79. Eastleach Martin (Clapper) Bridge.

80. Bibury (Foot) Bridge.

was granted for its repair. Even in those days a small property was held in trust to provide for its maintenance, and further land was bequeathed for this purpose in the sixteenth century.

Mr. W. J. Monk in his *History of Burford*, published in 1891, gives an extract from the *Gentleman's Magazine* for September 13th, 1797, which stated that 'in consequence of the heavy fall of rain the Windrush, a small river near this place, was so swelled in the course of the night as to carry away the bridge between this town (Burford) and Fulbrook.' According to the *Oxford Journal* for November 13th, 1829, 'the first stone of the intended alterations at Burford Bridge' had just been laid, and it thus appears that part of the bridge was rebuilt early in the nineteenth century. It was possibly the arch nearest the town, as this one appears to be of later construction than the others.

The Windrush joins the river Thames at New Bridge, which carries the road from Kingston Bagpuize to Stanlake. This bridge is illustrated and described in the *Ancient Bridges of the South of England*, but since that book was written further information has been found which helps to date the bridge. In the third volume of *Parochial Collections*, compiled by the Oxford Record Society, is the following note, under the heading of the parish of Stanlake:

'New Bridge built as 'tis said (or at least repaired) temp. Hen. 6 (1422-1461) by John Golafre whom some style Esquire, and some Kt. But this bridge being fallen into decay about 2. Ed. IV (1463) severall complaints were put up by the men of Kingston

Bakepuze and Stanlake for to have it repaired—whereupon one Thom. Briggs that lived in an Hermitage at the end of this bridge next to Stanlake obtained licence to require the good will and favour of passengers that came that way and of the neighbouring villages, so that money being then collected the bridge was repaired in good sort.'

Robert Plot in his *Natural History of Oxfordshire*, written in 1677, mentions an unusual custom which is probably related to this hermitage. He states that ' the Parson in the Procession about holy Thursday reads a Gospel at a Barrels head in the Cellar of the Chequer Inn, where some say there was formerly a Hermitage; others, that there was anciently a Cross at which they read a Gospel in former times, over which now the house, and particularly the cellar being built, they are forced to perform it in manner as above.'

THE RIVER LEACH AND THE RIVER COLN

The river Leach rises near Northleach in the Cotswolds and, flowing in a south-easterly direction, joins the river Thames about three miles below Lechlade. Its bridges are mostly small, and the only unusual one is the clapper bridge at Eastleach Martin which is shown in Fig. 79.

The river Coln takes a course more or less parallel to that of the Leach, but rises several miles farther north-west, at Brockhampton, about five miles east of Cheltenham. It is crossed by the Fosse Way at Fosse Bridge, a late eighteenth-century stone bridge with projecting keystones, and at Bibury by one of the same period which actually bears the date 1770. About 200 yards below the latter bridge is an attractive foot-bridge (Fig. 80), only 6 feet in overall width.

Leland recorded that the bridge at Fairford had four stone arches, but the present one was certainly not built before the middle of the eighteenth century.

THE RIVER CHURN AND THE AMPNEY BROOK

Both the Churn and the Ampney Brook, which join the Thames near Cricklade, also rise in the Cotswolds, the former at Seven Springs, about four miles south of Cheltenham. They are crossed by numerous bridges, mostly quite small, but several possibly date from the end of the eighteenth century.

INDEX OF BRIDGES

Abbot's (Lark), 109, 110
Abbot's (Tybourne), 144
Abridge, 136
Achelsford, 82
Acle (see Wey), 116
Alconbury, 98
Aldreath Causeway, 103
Alport, 28
Alton, 20
Anstey, 50
Arthingworth, 73
Ashbourne, 19
Ashford, 27
Ashfordby, 48
Ash Street, 127, 128
Asshton, 77
Aston (Dove), 21
Aston (Trent), 2
Astonfield, 17
Attlebridge, 119
Audley End, 104
Aylesbury, 150
Aylesham, 115
Aylestone, 44

Bakewell, 27
Baldwin, 149
Banbury, 153
Barford (Ise), 74
Barford (Ouse), 92
Barnwell, 77
Barrow-upon-Soar, 50, 51
Baslow, 25, 26, 27
Bawtry, 35
Beccles, 124
Bedford, 89, 90, 91, 92
Beggars, 149
Belgrave, 47
Belper, 30
Berkhampstead, 148
Bibury, 158
Bickerston, 117
Biddenham, 89
Biggleswade, 94
Billing, 72, 73
Birmingham, 38

Bishop (Norwich), 119, 121, 122
Bishop's (West Rasen), 56
Blackfriars, 119, 120
Blackwater, 132
Blatherwycke, 78
Blaydon, 155
Blunham, 94
Blyford, 124
Blyth (Ryton), 36
Blythe (Blithe), 16
Blythford, 16
Bole, 41
Boston, 60, 61
Bottesford, 52
Bourn, 126, 127
Bourton, 156
Bourton-on-the-Water, 156
Bow (Leicester), 44-47
Bow (Stratford), 142
Brackley, 83
Brampton, 73
Brancey, 76
Brandon, 112, 113
Brent, 147
Brentford, 147
Brigstock, 76
Broadeng, 64
Bromham, 89, 90
Buckingham, 82, 83
Bungay, 123, 124
Buntingford, 138
Burford, 156, 157
Burghley House, 65
Burnhurst Mill, 16
Burton, 48
Burton upon Trent, 5
Bury St. Edmunds, 109, 110
Buxton, 115

Caldecott, 62
Calver, 25
Cambridge, 104-107
Carrow (Norwich), 122

Castle (Buckingham), 83
Castle (Hertford), 138
Castle Acre, 114
Castle Hedingham, 129
Castleford, 66
Cattawade, 128, 129
Cavendish, 7, 8, 26
Chappell, 138
Charwelton, 152
Chatsworth, 26
Cheddleston, 20
Chelmsford, 134
Chelsworth, 127
Cheney, 11
Chetwynd, 42
Childe, 82
Chislehampton, 151
Church (Empingham), 65
Church End, 133
Clare, 107
Clayhythe, 108
Claypole, 57, 58
Cock (Ashop), 23
Cock (Waveney), 124
Coggeshall, 131, 132
Colchester, 130, 131
Coldwall, 18
Coleshill, 39
Collyweston, 63
Colnbrook, 148
Colton Mill, 3
Coslany, 119, 120
Cossington, 50
Cotes, 51
Cow, 138
Crasswell, 16
Cringleford, 117
Cromford, 29
Cropredy, 152
Crowdecote, 17
Crowland, 67, 68
Crowthorpe, 78
Curdworth, 39

Darlaston, 1
Darley, 29

161

INDEX OF BRIDGES

Deeping Gate, 67
Denford, 75
Derby, 31, 32, 33
Derwent, 26
Desborough, 74
Devon, 53
Digbeth, 38
Ditchford, 74, 75
Dorchester, 152
Dove, 20, 21
Draycott, 20
Duddington, 63
Duffield, 31
Dukes' Palace, 122
Dunham, 15

Eamont, 22
Earith, 103
East (Colchester), 130
East (Oxford), 154
East Gate, 109
Eastleach Martin, 158
East Lexham, 114
East Retford, 35
Eddingham, 65
Edenham, 68
Edgeware, 146
Eggington, 22
Elford, 42
Ellaston, 19
Ely, 108
Emley, 153
Empingham, 65
Enderby Mill, 44
Enslow, 153
Essendine, 68
Essex, 2
Eustace, 142
Euston, 111, 112
Evenley, 82
Everdon, 71, 72
Exeter (Derby), 32, 33

Fairford, 159
Fakenham, 111
Farcet, 81
Fazeley, 40, 41
Felmersham, 88
Fenny Stratford, 86
Fielden, 41
Fillyford, 28
Fingringhoe, 131

Fleet, 143
Fleming's, 52
Flitcham, 114
Folly, 138
Ford, 139
Fosdyke, 69
Fosse (Coln), 158
Fosse (Foss Dyke), 58
Fotheringhay, 78, 79
Foundry, 122
Framwellgate, 22
Froggatt, 24, 25
Frogmire, 45
Ful, 133
Furnace End, 40
Fye, 119, 120

Gainsborough, 15
Gallows, 150
Garrett Hostel, 107
Geddington, 74
Gedling, 13
George, 64
Girtford, 94
Glandford, 56, 57
Glutton, 17
Godmanchester, 98
Gosford, 154
Gote, 59
Great (Cambridge), 105, 106
Great Barford, 92, 93
Great Bowden, 62
Great Casterton, 65, 66
Great Castleford, 66
Great Dunmow, 133, 134
Great Oxenden, 73
Great Yarmouth, 123
Grindleford, 24

Haddon, 28
Hadleigh, 128
Hail, 97
Halstead Rural, 135
Hammerton, 98
Hampton in Arden, 39
Hanging, 19
Hanwell, 147
Harford, 117, 118
Harrington, 8
Harringworth, 62
Harris, 41

Harrold, 87, 88
Hartington, 17
Hay, 15
Hayward, 152
Haywood, 2
Hazelford, 24
Heacham, 114
Heigham, 116, 117
Hemlingford, 40
Hertford, 138
Hethebeth, 9-13
Hey, 20
Heyford, 153
High (Blackwater), 133
High (Ouse), 103
High (Trent), 3
High (Witham), 59
Hilgay, 113
Holborn, 143
Holland Causeway, 70
Holmans, 150
Holme, 27
Honiton, 112
Hopwas, 42
Houghton St. Giles, 115
Hoveton St. John, 115
Hoybel, 9
Hunstanton, 114
Huntingdon, 97-101
Hunton, 148
Huntspill, 13
Hythe (Colchester), 130, 131

Ickford, 150
Icklingham, 110
Incelland, 112
Ipswich, 126
Irthlingborough, 75
Islip, 154
Ixworth, 111

Kate's, 69
Kegworth, 51
Kelham, 14, 15
Kentford, 110
Kettleby, 48
Ketton, 65
King William's, 50
King's, 3
King's Bromley, 3

INDEX OF BRIDGES 163

King's College (Cambridge), 106, 107
King's Langley, 148
King's Lynn, 114
Knight's (London), 145

Lady, 41
Lammas, 64
Langford, 93
Langham, 43
Latchleys Manor House, 135
Lathbury, 85
Leadmill, 24
Leen, 12
Leicester, 44-47
Leighton Buzzard, 86
Lenwade, 119
Lesser Bow, 46
Lewin, 49
Lilford, 76, 77
Lincoln, 59
Linslade, 86
Lion's, 43
Little Bowden, 62
Little Chester, 31
Little Chesterford, 103, 104
Little North, 45
Littleport, 108
Lolham, 66
London (Buckingham), 83
London Colney, 148
Long (Blackwater), 130, 131
Long (Nene), 73
Longe, 148
Longnor, 17
Lord's, 83
Loughton, 136
Louth, 56
Lower Leigh, 16
Lower Slaughter, 156
Luton, 137

Magdalen, 154, 155
Maldon, 133
Mansfield, 34
Manthorpe, 68
Manton, 65
Mapleton, 18
Marble Works, 27

March, 81
Marham Claybrigge, 60
Market Deeping, 67
Market Harborough, 61, 62
Markham, 53, 54
Marston, 57
Matlock, 29
Mattersay, 35
Mayfield, 19
Mayton, 116
Medborne, 62
Melancholy Walk, 65
Melford, 111
Melton Mowbray, 48
Merriel, 34
Middle, 62
Middleton, 62
Mill (Hertford), 138
Milldale, 17
Monk's, 22
Moulsham, 134, 135
Moulton, 110
Mountsorrel, 50
Mulle, 148
Mundford, 113
Muskham, 14, 15
Mytham, 24

Narborough (Nar), 114
Narborough (Soar), 43
Nell, 153
New (Colne), 130
New (Derwent), 25
New (Norwich), 120
New (Ryton), 36
New (Thames), 157, 158
New (Windrush), 156
Newark, 14, 53
Newbottle, 73
Newcastle-under-Lyme, 1
Newport Pagnell, 85
Newton Flotman, 118
Nine Days' Wonder, 49
North (Colchester), 130
North (Leicester), 44, 45
North (Newport Pagnell), 85
North (Oundle), 77
North (Shefford), 93
North (Towcester), 85
Northall, 85

Northampton, 72
Northdyke, 55
Norwich, 119-123
Nottingham, 8-13
Nuns' (Alconbury), 97, 98
Nuns' (Thet), 112

Oakamore, 20
Okeover, 18
Old (Belgrave), 47
Old Lakenham, 118
Oldehe, 58
Ollerton, 33
Olney, 86
Omanne, 150
One Arch, 26
Oundle, 77, 78
Outwell, 81
Oxford, 154

Padbury, 83
Park, 133, 134
Passenham, 84
Passingford, 136
Perry, 38
Peterborough, 80
Pilton, 76
Pleshey Castle, 135
Polesworth, 41
Prickwillow, 110

Queniborough, 49

Radwell, 88
Ramsey, 81
Rearsby, 49
Rocester, 20
Rockingham, 62
Rowsley, 26, 27
Rushford, 111
Ruston, 74
Ryhall, 66

St. George's, 120
St. Ives, 102
St. John's College, 106, 107
St. Marteyns, 121
St. Mary (Tame), 41
St. Mary's (Derwent), 32, 33
St. Mary's (Dove), 18
St. Miles, 120

INDEX OF BRIDGES

St. Neots, 95, 96, 97
St. Olaves, 124
St. Peter's, 72
St. Sundays, 44, 45, 46
St. Thomas, 72
Salter's, 42
Sandham, 50
Sandon, 2
Sandy-brook, 19
Sapiston, 111
Schole, 123
School House, 19
Selford, 112
Sheepwash, 27
Shefford (Colne), 148
Shefford (Ivel), 93
Sheriffs, 83
Sherrington, 86
Sighs (Cambridge), 107
Sleaford, 60
Small (Cambridge), 105-107
Smite, 53
Snape, 126
South (King's Lynn), 114
South (Newport Pagnell), 85
South (Northampton), 72
South (Shefford), 93
South Collingham, 15
South Ferriby, 55, 57
Spalding, 68
Spaldwick, 98
Spitel (Eton), 149
Spetil (Aylesbury), 150
Springfield, 134
Stableford, 37
Stafford, 37
Stafford (Ouse), 88
Stamford, 64, 65
Stanborough, 137
Standen, 139
Stanford-upon-Soar, 51
Stanne, 150
Stanstead, 141
Stanway, 131
Stanwell, 149
Stapleford, 48
Staughton Highway, 97
Stickledon, 146
Stoke, 126
Stoke Ferry, 113

Stoken, 137
Stone (Piccadilly), 144
Stoney, 43
Stony Stratford, 82, 84
Stratford (Blackwater), 132
Stratford (Deben), 126
Stratford (Lea), 142
Sudbury, 127
Surfleet, 69
Sutton, 94
Swarkeston, 6, 7
Syleham, 123
Syston, 49

Tallington, 66
Tame, 37, 38
Tamworth, 41
Tattershall, 59, 60
Tempsford, 95
Thame, 150
Thele, 137, 141
Thetford, 111, 112
Thornborough, 83, 84
Thornton, 84
Thorpe, 18
Thoty, 82
Thrapston, 75, 76
Thrussington, 49
Tickford, 86
Till, 58
Toft, 69
Tolleshunt D'Arcy, 135
Toppesfield, 128
Torksey, 58
Totis, 83
Towcester, 85
Trafford, 152
Trent (Newark), 53, 54
Trent (Nottingham), 11, 12, 13
Trinity (Crowland), 67, 68
Trowse, 118
Turvey, 87
Turweston, 82
Tutbury, 21, 22
Twiford, 82
Twyford, 35

Uffington, 66
Upper Leigh, 16

Upper Tean, 20
Utterby, 56
Uxbridge, 148

Viators, 17

Wadesmill, 139, 140
Wakerley, 63, 64
Walcott, 79
Wall, 150
Walpole, 124
Waltham, 141, 142
Walton (Thame), 150
Walton (Trent), 1, 2
Wansford, 79, 80
Ware, 137, 140
Warkworth, 22
Wash Brook, 94
Waterend, 148
Water-furlong, 64
Water Orton, 38, 39
Watton at Stone, 138
Welland, 62
Welland (Stamford), 64, 65
West (Leicester), 44, 45, 46
West (Northampton), 72
West Acre, 114
West Deeping, 66
West Lexham, 114
West Rasen, 56
Weston, 2
Wey, 116
Whatstandwell, 30
Wheatley, 151
Whitefriars, 119, 121
Whittlesford, 104
Whitton Mill, 72
Wichnor, 4, 5
Windsor, 149
Windy Arbour, 17
Wisbech, 81
Wistow, 81
Wiveton, 115
Wolseley, 3
Woodford, 136, 137
Woodstock, 155

Yorkshire, 23, 24
Yoxall, 3, 4

Zouch, 51

www.ingramcontent.com/pod-product-compliance
Lightning Source LLC
Chambersburg PA
CBHW031426150426
43191CB00006B/416